GCSE
Biology

Adrian Schmit, Jeremy Pollard

HODDER
EDUCATION
AN HACHETTE UK COMPANY

This material has been endorsed by WJEC and offers high quality support for the delivery of WJEC qualifications. While this material has been through a quality assurance process, all responsibility for the content remains with the publisher.

WJEC past paper questions are reproduced by permission of WJEC.

Although every effort has been made to ensure that website addresses are correct at time of going to press, Hodder Education cannot be held responsible for the content of any website mentioned in this book. It is sometimes possible to find a relocated web page by typing in the address of the home page for a website in the URL window of your browser.

Hachette UK's policy is to use papers that are natural, renewable and recyclable products and made from wood grown in well-managed forests and other controlled sources. The logging and manufacturing processes are expected to conform to the environmental regulations of the country of origin.

Orders: please contact Hachette UK Distribution, Hely Hutchinson Centre, Milton Road, Didcot, Oxfordshire, OX11 7HH. Telephone: +44 (0)1235 827827. Email education@hachette.co.uk. Lines are open from 9 a.m. to 5 p.m., Monday to Friday. You can also order through our website: www.hoddereducation.co.uk

ISBN: 978 1 4718 6871 9

© Adrian Schmit and Jeremy Pollard 2016

First published in 2016 by

Hodder Education,

An Hachette UK Company

Carmelite House

50 Victoria Embankment

London EC4Y 0DZ

www.hoddereducation.co.uk

Impression number   10 9 8 7 6

Year   2022

Cover photo © Steve Taylor ARPS / Alamy Stock Photo
Typeset in India by Aptara Inc.

Printed by CPI Group (UK) Ltd, Croydon CR0 4YY
A catalogue record for this title is available from the British Library.

Contents

Get the most from this book

Welcome to the WJEC GCSE Biology Student Book.

This book covers all of the Foundation and Higher-tier content for the 2016 WJEC GCSE Biology specification.

The following features have been included to help you get the most from this book.

 Specification coverage

Check that you are covering all the required content for your course, with specification references and a brief overview of each chapter.

Key terms

Important words and concepts are highlighted in the text and clearly explained for you in the margin.

Activity

These activities usually involve the use of second-hand data that could not be obtained in the school laboratory, along with questions that will test your scientific enquiry skills.

Discussion point

These are questions that could be answered by individuals, but that benefit from discussion with your teacher or others in your class. In such cases there are usually a variety of opinions or possible answers to explore.

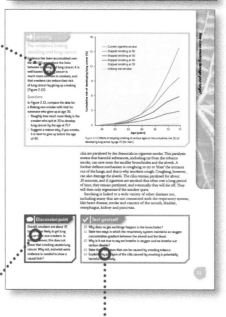

Practical

These practical-based activities will help consolidate your learning and test your practical skills.

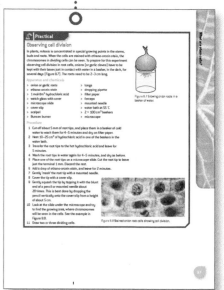

Test yourself

These short questions, found throughout each chapter, allow you to check your understanding as you progress through a topic.

Worked example

Examples of questions and calculations that feature full workings and sample answers.

Chapter review questions

You will find practice questions at the end of every chapter. These follow the style of the different types of questions you might see in your examination and have marks allocated to each question part.

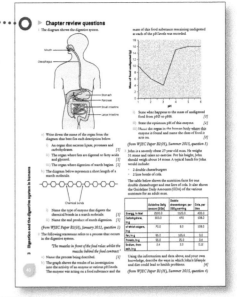

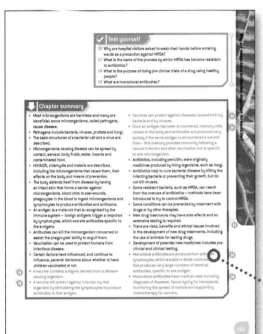

Chapter summary

This provides an overview of everything you have covered in a chapter and is a useful tool for checking your progress and for revision.

Some material in this book is only required for students taking the Higher-tier examination. This content is clearly marked by the Higher **H** icon.

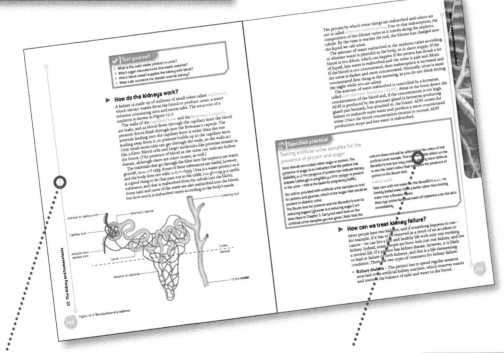

Most of the content in this book is suitable for all students. However, some topics should only be studied by those taking GCSE Biology. This content is clearly marked with a green line.

Specified practical

WJEC's specified practicals are clearly highlighted.

Answers

Answers for all questions and activities in this book can be found online at: www.hoddereducation.co.uk/wjecgcsebiology

Cells and movement across cell membranes

🏠 | **Specification coverage**

This chapter covers the GCSE Biology specification section
1.1 Cells and movement across membranes and GCSE Science
(Double Award) specification section 1.1 Cells and movement
across membranes.

It covers the structure and function of cells, how they
transport materials and some metabolic processes that occur
within them.

▶ What are cells?

Cells are now known to be the basic 'unit' of all living things. Cells
were first seen through a microscope and described by the famous
scientist Robert Hooke in 1665 (Figure 1.1), but at that time he had
no idea that cells were found in all living things. That idea, which
formed part of what is known as the cell theory, was first suggested
by German scientists Theodor Schwann (working on animals) and
Matthias Schleiden (working on plants) in the 1830s.

The cell theory that Schwann and Schleiden proposed is still
the basis of cell theory today, although it has been developed as we
have come to know more about cells.

Today's cell theory states that:

Figure 1.1 An early microscope used by
Robert Hooke to discover cells.

▷ All living organisms are composed of cells. They may be
  unicellular (one celled) or multicellular (many celled).
▷ The cell is the basic 'unit' of life.
▷ Cells are formed from pre-existing cells during cell division.
▷ Energy flow occurs within cells (enabling the chemical reactions
  that make up life to take place).
▷ Hereditary information (deoxyribonucleic acid, DNA) is passed
  on from cell to cell when cell division occurs.
▷ All cells have the same basic chemical composition.

Although all cells have features in common, there are also differences
between different types of cell. Some of those differences allow
scientists to classify cells as either animal cells or plant cells.

▶ Plant and animal cells

All cells in both plants and animals have certain features in common:

▷ They all have cytoplasm, a sort of 'living jelly', where most of the
  chemical reactions that make up life go on.
▷ The cytoplasm is always surrounded by a cell membrane, which
  controls what enters and leaves the cell.

- They have a nucleus, which contains DNA, the chemical which controls the cell's activities.
- They contain mitochondria (singular: mitochondrion), which are the structures that carry out aerobic respiration, supplying cells with energy.

Plant cells can be distinguished from animal cells, because they have some features that are not seen in animal cells. These are:

- a cell wall, made of cellulose, which surrounds all plant cells
- a large, permanent central vacuole, which is a space filled with liquid cell sap
- chloroplasts, which absorb the light plants need to make their food by photosynthesis − chloroplasts are not found in *all* plant cells, but they are never found in animal cells.

Figure 1.2 shows examples of plant and animal cells, showing the differences between them.

Figure 1.2 Animal cell (left) and plant cell (right) showing differences in structure.

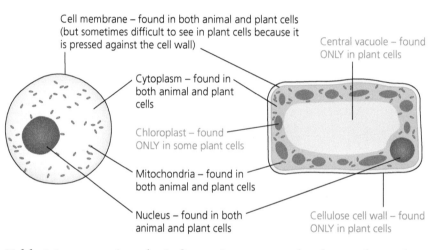

Cell membrane – found in both animal and plant cells (but sometimes difficult to see in plant cells because it is pressed against the cell wall)

Cytoplasm – found in both animal and plant cells

Central vacuole – found ONLY in plant cells

Chloroplast – found ONLY in some plant cells

Mitochondria – found in both animal and plant cells

Nucleus – found in both animal and plant cells

Cellulose cell wall – found ONLY in plant cells

Table 1.1 summarises the information you need to know about the structure of cells.

Table 1.1 A summary of cell structure.

Organelle	Where found	Function
Nucleus	All cells	Contains DNA, which controls the cell's activities
Cell membrane	All cells	Controls what enters and leaves the cell
Cytoplasm	All cells	Forms the bulk of the cell and is where most of the chemical reactions occur
Mitochondria	All cells	Provide energy by carrying out aerobic respiration
Chloroplasts	Some plant cells	Absorb light for photosynthesis
Cell wall	Plant cells	Supports the cell
Vacuole	Plant cells	Filled with a solution of nutrients including glucose, amino acids and salts

Test yourself

1 State three features that are found in both animal and plant cells.
2 State three features that are found in plant cells but not in animal cells.
3 Muscle cells have a lot of mitochondria. Suggest a reason for this.
4 What is the function of the cell wall in plant cells?
5 Suggest a reason why many plant cells do not contain chloroplasts.

Observing cells with a microscope

Throughout history since the time of Robert Hooke, scientists have used light microscopes to observe cells. There are now more powerful types of microscope available (for example, electron microscopes), which have allowed us to see the detailed structure of cell organelles, but these are complex and expensive and the light microscope is still by far the most widely used type of microscope. The parts of a typical light microscope are shown in Figure 1.3.

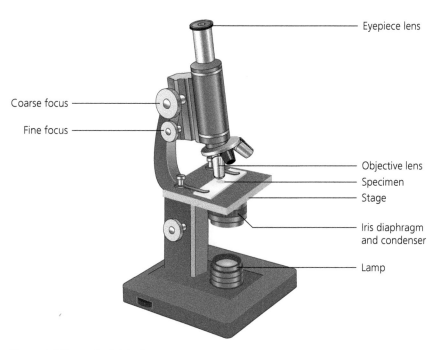

Figure 1.3 Parts of a light microscope.

The functions of the parts of a light microscope are as follows:

▶ The **eyepiece lens** is of fixed magnification, although it is possible to exchange it with a lens of a different magnification.
▶ The **objective lenses** are of different magnifying power and are interchangeable. They are the ones used to adjust the magnification of the image that you see down the microscope.
▶ The **stage** is where the microscope slide you are observing is placed, with clips to hold it in place.
▶ Below the stage is a part that is usually made of up of two components – an **iris diaphragm**, which can be opened or closed to adjust the amount of light entering the objective lens, and (sometimes) a **condenser**, which concentrates the light into a beam directed precisely into the objective lens.
▶ At the base of the microscope is a **lamp**, although some older microscopes may have a mirror, which is used in conjunction with a separate lamp to shine light through the condenser and iris diaphragm.
▶ The microscope is focused using two focus controls. The **coarse focus** control is used to get the image roughly into focus using the lowest-power objective, and then the **fine focus** control is used to fine tune the image and make it as clear as possible.
▶ Microscope **slides** hold thin specimens or sections, which may be stained using a variety of dyes so that structures can be seen more clearly.

Microscope drawings

The purpose of a scientific drawing of a microscope specimen is to show it accurately (correct shapes and proportions) and as clearly

Examination of animal and plant cells using a microscope, and production of scientific labelled diagrams

In this practical you will look at animal cells from your own body and plant cells from an onion. As onions are formed underground and are not exposed to light, their cells do not have chloroplasts. It is also unlikely that the microscope you use will be powerful enough to see mitochondria.

Animal cells

Apparatus

> cotton wool buds/interdental sticks
> microscope slide
> cover slip
> microscope
> methylene blue stain
> filter paper

Procedure

1 Gently scrape the inside of your cheek with a cotton wool bud or interdental stick (see Figure 1.4).
2 Smear the saliva from the bud gently onto your microscope slide.
3 Add a drop or two of water to the part of the slide you smeared.
4 Place the cover slip onto the slide. Place one edge on the slide and then gently lower the cover slip down.
5 Add a drop of methylene blue dye near one edge of the cover slip, on the microscope slide.
6 Draw the dye under the cover slip by putting the filter paper next to the opposite edge of the cover slip to the dye.
7 Leave a few minutes for the dye to stain the cells, then observe the slide under the microscope.
8 Make a drawing of three cells.

Plant cells

Apparatus

> onion
> scalpel
> forceps
> microscope slide

> cover slip
> methylene blue/iodine stain
> microscope

Procedure

1 Split and pull the onion apart into layers.
2 With the scalpel, make a square cut part way through an onion segment. Make sure that this is on the side of the segment that was towards the inside of the onion, and that the square cut is smaller than the cover slip.
3 Using the forceps, carefully peel the inner epidermal cell layer away from the onion (Figure 1.5). The epidermal layer (or epidermis) is a 'skin' one cell thick on the inside and outside of each of the layers of the onion. Use the inner side of the onion layer for this procedure.
4 Place a drop of methylene blue or iodine stain onto the centre of the slide.
5 Gently lay the sheet of epidermis onto the drop of dye. Try to avoid trapping air bubbles underneath the tissue.
6 Lay a cover slip over the tissue.
7 Leave a few minutes for the dye to stain the cells, then observe the slide under the microscope.
8 Make a drawing of a group of no more than four cells.

Figure 1.4 Sampling cheek cells.

Figure 1.5 Removing the epidermis from an onion.

as possible. You do not have to be a wonderful artist to produce a good scientific drawing, but you do have to be observant and neat. There are certain 'rules' about producing scientific diagrams:

1 Always use a sharp pencil.
2 Make sure lines are thin, clear and do not overlap where they join.
3 Do not shade parts of your drawing, unless it is absolutely essential in order to clearly distinguish structures.
4 Always use a ruler for labelling lines. The lines should never cross each other.
5 Do not label on the drawing itself – keep the labels outside the drawing.
6 Make sure the proportions of the drawing are correct. If a structure is twice as wide as it is long, then the drawing of it should be, too.

Specialised cells

Just like whole organisms, cells have evolved over time to become specialised for their particular 'jobs' (Figure 1.6). This can sometimes result in cells that look very different from the examples in Figure 1.2.

Newly formed cells in different tissues look very similar to one another, no matter what tissue they come from. As they grow and mature, however, they gradually develop the specialisations that suit their function. This process of change is called differentiation. Let's take a red blood cell as an example. The cell is formed as a basic animal cell, very similar to that shown in Figure 1.2. Over a period of about two days, the cell gradually loses its nucleus and organelles, forms haemoglobin (which is the pigment needed to carry oxygen) and acquires its characteristic biconcave shape to become a fully formed red blood cell.

Figure 1.6 Specialised cells.

  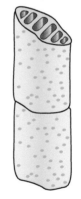

Sperm cell
The cell has very little cytoplasm and a tail, to help it swim fast towards the egg

Red blood cells
The cells have lost their nuclei and have become packed with a red pigment, haemoglobin, which carries oxygen around the body

Xylem cells
The xylem cells form tubes which carry water up a plant, and also strengthen it. To do this, the cells have perforated end walls, the cell wall is very thick, and the cytoplasm has died off to leave a hollow tube

▶ How are cells organised into a whole body?

During the development of an animal or plant, the cells organise themselves into groups called tissues. Different tissues are grouped together to form organs, and the organs may link up to form organ systems (the organs may not be physically linked – they may just have linked functions). All of the organ systems working together form a whole animal or plant – which is known as an organism. Definitions and examples of the different levels of organisation are shown in Table 1.2 on the next page. Note that the term 'organism' does not actually imply that organ systems are present – some living organisms consist of only one cell.

Table 1.2 Levels of organisation in the structure of living things.

Level of organisation	Definition	Examples
Tissue	A group of similar cells with similar functions	Bone, muscle, blood, xylem, epidermis
Organ	A collection of two or more tissues that perform specific functions	Kidney, brain, heart, leaf, flower
Organ system	A collection of several organs that work together	Digestive system, nervous system, respiratory system, shoot system, root system
Organism	A whole animal or plant	Cat, elephant, human, rose bush, oak tree

Bone is an example of a tissue. It is made up of two types of similar cells, which together form and maintain bones. Note that 'a bone' is an organ, as it consists of both bone tissue and blood.

A leaf is another example of an organ. It has several different types of cell which perform different functions, all linked to the production of food by photosynthesis (Figure 1.7).

Figure 1.7 The tissues in a leaf.

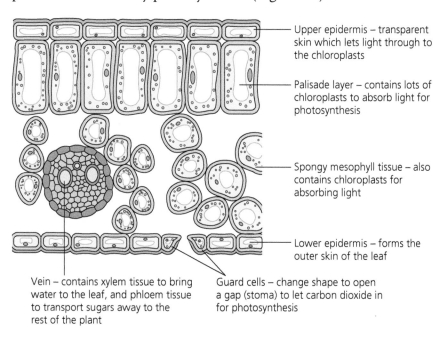

Upper epidermis – transparent skin which lets light through to the chloroplasts

Palisade layer – contains lots of chloroplasts to absorb light for photosynthesis

Spongy mesophyll tissue – also contains chloroplasts for absorbing light

Lower epidermis – forms the outer skin of the leaf

Vein – contains xylem tissue to bring water to the leaf, and phloem tissue to transport sugars away to the rest of the plant

Guard cells – change shape to open a gap (stoma) to let carbon dioxide in for photosynthesis

The digestive system is an example of an organ system. It consists of a number of organs (including the stomach, small intestine, liver and pancreas) that work together to digest and absorb nutrients.

▶ Movement into and out of cells

In order to get into and out of cells, substances have to get through the cell membrane. The cell membrane is selectively permeable, which means it lets some molecules through but not others.

Sometimes, you see the cell membrane referred to as 'partially permeable' rather than 'selectively permeable'. Don't worry – it's the same thing. In general, large molecules cannot get through the membrane, but smaller molecules can. Whether they actually do get through, which way they travel, and how quickly, depends upon a number of factors, as we shall see. There are three processes by which substances move through membranes:

▶ **diffusion**, when particles sort of 'drift' through the membrane
▶ **osmosis**, which is a special case of diffusion, involving water only
▶ **active transport**, when particles are actively 'pumped' through the membrane in a particular direction.

The statements in the list above are not full definitions. Those will be given later, as we consider each of these processes in detail.

Diffusion

Diffusion is the spreading of particles from an area of higher concentration to an area of lower concentration, as a result of random movement. We say the particles move down a **concentration gradient** (Figure 1.8).

Diffusion is a natural process that results from the fact that all particles are constantly in motion. It is called a **passive process**, because it does not require an input of energy. The movement is random – there is nothing pushing the particles and they cannot possibly 'know' in which direction they are heading. The particles will move in all directions, yet the *overall* (net) movement is always from an area of high concentration to an area of low concentration.

Two of the most important substances that enter and leave cells by diffusion are oxygen, which is needed for respiration, and carbon dioxide, which is a waste product of that process. The speed of diffusion can be increased by increasing the temperature, because that makes the particles move faster, or by increasing the concentration gradient (the difference between the high and low concentrations).

Osmosis

Osmosis is a specific type of diffusion. It is the diffusion of **water molecules** through a **selectively permeable membrane**. Diffusion of any other substance through a selectively permeable

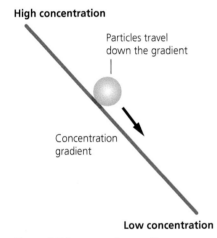

High concentration

Particles travel down the gradient

Concentration gradient

Low concentration

Figure 1.8 Concentration gradient.

🧪 Practical

Modelling diffusion

Procedure

1 Place about 10 marbles in a group on the laboratory bench. Ensure that they stay in a group and do not roll apart. These represent molecules in a high concentration. The surrounding area, with no marbles, represents a low concentration.
2 Bring your fists down on the bench on either side of the group of marbles. This will provide the marbles with energy and they should move.

3 Observe how they travel.
4 You should find that the marbles spread out from the group. In other words, they move from an area of high concentration to an area of low concentration.

Questions

1 The marbles never remain in a group, they always spread out. Explain why this happens.
2 In what way(s) is this model an inaccurate way of representing the movement of particles?

How does the membrane affect diffusion?

Small molecules can get through the cell membrane, but large molecules cannot. In this experiment, you will be using starch (a large molecule), iodine (a small molecule) and Visking tubing, which is a sort of cellophane with similar properties to a cell membrane. It has pores in it that let only small molecules through. Iodine stains starch blue-black when it comes into contact with it.

Safety notes

Wear eye protection.

Apparatus

> boiling tube
> length of Visking tubing, knotted at one end
> dropping pipette
> elastic band
> iodine in potassium iodide solution
> 1% starch solution
> test-tube rack

Procedure

1 Set up the apparatus as shown in Figure 1.9. Fill the Visking tubing with starch solution using the dropping pipette. Be careful that no starch drips down the outside of the tubing.
2 Place the boiling tube in a test-tube rack and leave for about 10 minutes.
3 Observe the result.
4 Explain the colours that you see after 10 minutes inside and outside the Visking tubing.

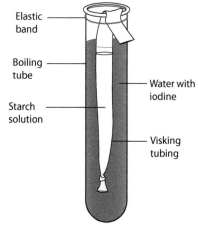

Figure 1.9 Apparatus for an experiment investigating how a membrane affects diffusion.

membrane is just called diffusion. Diffusion of water, but *not* through a membrane, is just diffusion. To be called osmosis, the process has to involve *both* water *and* a membrane. In osmosis, we say that water moves from a solution of low solute concentration (which has more water) to a solution of high solute concentration (which has less water), through a selectively permeable membrane. Notice that the substance (water) that is diffusing is still going down a concentration gradient. A concentrated solution of salt, for instance, would have a low 'concentration' of water, whereas a dilute solution would have a high 'concentration' of water. The movement of water happens because the membrane is permeable to water (that is, it lets it through), but not to the solute. The process of osmosis is shown in Figure 1.10.

All the molecules on both sides of the membrane are moving. Occasionally, a molecule hits a membrane 'pore'. Water molecules will go through but solute molecules will not. Because there is a higher proportion of water molecules in the dilute solution, more will travel from the dilute solution to the concentrated solution than the other way. Although water molecules move in both directions, there is net movement from the dilute to the more concentrated solution.

If the concentrations of the solutions on either side of the membrane are the same, then overall an equal number of water molecules travel in each direction – we say that such solutions are **in equilibrium**.

Figure 1.10 The process of osmosis.

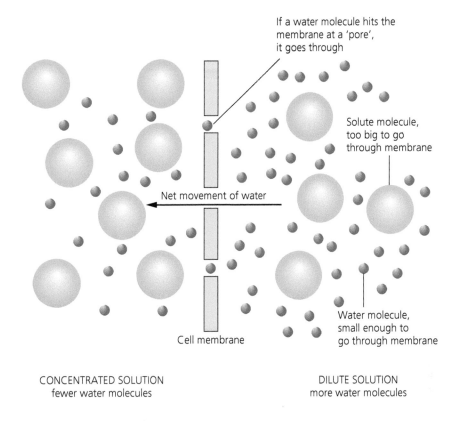

Figure 1.10 The process of osmosis.

If a water molecule hits the membrane at a 'pore', it goes through

Solute molecule, too big to go through membrane

Net movement of water

Water molecule, small enough to go through membrane

Cell membrane

CONCENTRATED SOLUTION
fewer water molecules

DILUTE SOLUTION
more water molecules

Why is osmosis important?

Osmosis is important because too much or too little water inside cells can have disastrous effects. If an animal cell is put into a solution that is more dilute than its cytoplasm, water will go in by osmosis and the cell will burst. If a patient in hospital needs extra fluid, they are often put on to a 'saline drip'. Saline is a solution of salts at the same concentration as the blood. If just water was given, the blood would become too dilute and osmosis would make the blood cells burst.

Plant cells are not damaged by being put into water. They swell as water enters, but their cell wall stops them bursting. However, they can be damaged (as can animal cells) by being put into a concentrated solution. In this case, water leaves the cell by osmosis, and the cytoplasm collapses and shrinks. In plant cells, the cytoplasm pulls away from the cell wall, a condition known as plasmolysis (Figure 1.11). Plasmolysis can result in the death of the cell.

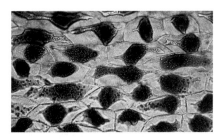

Figure 1.11 The plant cells have plasmolysed. Water has left the cells by osmosis and the cytoplasm has shrunk and pulled away from the cell wall.

Active transport

Diffusion, and the special form of it called osmosis, both transport substances down a concentration gradient. That is the 'natural' way for particles to move. Sometimes, though, cells need to get particles into or out of the cytoplasm against a concentration gradient. In other words, they have to be moved from an area of lower concentration to an area of higher concentration. This will not happen by diffusion, and in order to move the particles, the cell has to use energy to 'pump' the particles in the direction they need to go. As this type of transport requires an input of energy, it is called **active transport**.

Comparing active transport, diffusion and osmosis

Figure 1.12 shows the similarities and differences between the three cell transport processes.

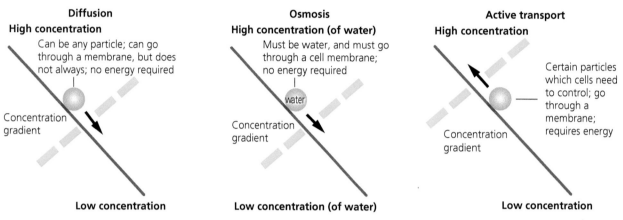

Figure 1.12 Comparison of diffusion, osmosis and active transport.

✔ Test yourself

6 Why is diffusion described as a passive process?
7 Why is active transport necessary?
8 When water evaporates, it spreads through the air by diffusion. Why would it be incorrect to call this osmosis?
9 What conditions cause plasmolysis in plant cells?
10 Why is it important that any fluid put into the bloodstream has the same concentration as the blood?

► How are the activities of a cell controlled?

All of the activities of a cell depend on chemical reactions. It has been estimated that 10 million reactions occur in a typical cell every second. These reactions are controlled by special molecules called **enzymes**. Which enzymes are produced in cells is controlled by another molecule, **deoxyribonucleic acid (DNA)**, which is found in the cell nucleus.

Enzymes

Enzymes are protein molecules that act as **catalysts**. A catalyst is something that speeds up a chemical reaction. It doesn't react itself, it simply causes the reaction it catalyses to go faster. Here are some important facts about enzymes:

▸ Enzymes act as catalysts, speeding up chemical reactions.
▸ The enzyme is unchanged by the reaction it catalyses.
▸ Enzymes are specific, which means that a certain enzyme will only catalyse one reaction or one type of reaction.
▸ Enzymes work better as temperature increases, but if the temperature gets too high they are destroyed (denatured).

▶ Different enzymes are denatured at different temperatures.

▶ Enzymes work best at a particular 'optimum pH' value, which is different for different enzymes.

▶ The chemicals on which enzymes work are called substrates. In order to catalyse a reaction, the enzyme has to 'lock together' with its substrate. The shapes of the enzyme and substrate must match, so that they fit together like a lock and key. That is why enzymes are specific – they can only work with substances that fit with their particular shape.

The action of this 'lock and key' model is shown in Figure 1.13.

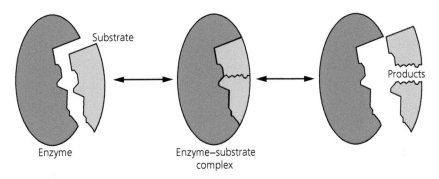

Figure 1.13 The 'lock and key' model of enzyme action. Note that in some reactions an enzyme catalyses the breakdown of a substrate into two or more products, while in others an enzyme causes two or more substrate molecules to join to make one product molecule.

The effect of temperature and pH on enzymes

Warming an enzyme actually makes it work faster at first, because the enzyme and substrate molecules move around faster and so meet and join together more often. But at higher temperatures the enzyme stops working altogether.

You can see from the 'lock and key' model in Figure 1.13 that the shape of the enzyme molecule is important if it is to work. The reason that enzymes won't work if they are at too high a temperature or at the wrong pH is because in these conditions their shape is altered, so that they no longer fit the substrate.

The part of an enzyme which binds to a substrate is called the active site and it is held in shape by chemical bonds. High temperatures and unsuitable pH conditions can break these bonds. This changes the shape of the active site so that the substrate molecule will no longer fit. The enzyme no longer works, and is said to be denatured. The actual temperature that denatures enzymes is different for different enzymes. Some enzymes start to denature at about 40 °C, most denature at around 60 °C, and boiling is fairly certain to denature an enzyme. A very small number of enzymes, mostly found in bacteria, have been found to tolerate temperatures as high as 110 °C.

When scientists want to test if a certain substance is acting as an enzyme, they do a controlled experiment in which a boiled sample of the substance is used in place of an un-boiled sample. If the reaction does not then occur, it is assumed that the substance acts as an enzyme.

The effect of temperature on enzyme action is shown in Figure 1.14, and the effect of pH is shown in Figure 1.15.

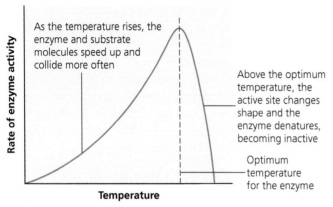

Figure 1.14 The effect of temperature on enzymes.

As the temperature rises, the enzyme and substrate molecules speed up and collide more often

Above the optimum temperature, the active site changes shape and the enzyme denatures, becoming inactive

Optimum temperature for the enzyme

Rate of enzyme activity

Temperature

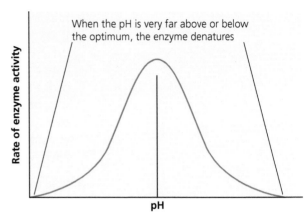

Figure 1.15 The effect of pH on enzymes.

When the pH is very far above or below the optimum, the enzyme denatures

Rate of enzyme activity

pH

Specified practical

Investigation into factors affecting enzymes

What is the best temperature at which to wash your clothes?

Enzymes are used for many commercial and industrial purposes, including 'biological' washing detergents. Many of the hardest stains to remove are mainly lipid (such as oils and butter) or protein (for example, blood and grass). The inclusion of enzymes that break down lipids (lipases) and proteins (proteases) in biological washing detergents helps to remove these stains (Figure 1.16). The enzymes used are more resistant to high temperatures than most, but can still be denatured.

Egg yolk is a good stain to test because egg yolk consists mostly of protein with a small amount of lipid. Design and carry out an experiment to test what temperature is best to use with a given brand of biological detergent (in any form). When designing the experiment, consider the following:

> How are you going to 'measure' how successful the detergent has been?
> How are you going to make the test fair?
> How are you going to make sure that you are measuring the effect of the detergent, rather than just the temperature of the water it is in?
> The experiment will not be valid unless the detergent is maintained at approximately its designated temperature throughout the experiment.
> You need to do a risk assessment of your experiment and ask your teacher to check it.

Figure 1.16 Biological detergents contain a mixture of enzymes to break down stains.

Analysing and evaluating your experiment

1   What is your conclusion from the experiment?
2   How strong is the evidence for your conclusion? Explain your answer.
3   If you could re-design your experiment, is there anything you would change?
4   What other factors, apart from the effectiveness of stain removal, might influence a decision about what temperature to use for your wash?
5   Explain why enzymes allow washing at a lower temperature than non-biological detergents.

✔ Test yourself

11 Define the term 'catalyst'.
12 Why would it be incorrect to say that an enzyme reacts with a substrate?
13 Enzymes are specific – that is, they will only work with one type of substrate. Explain this using the 'lock and key' theory.
14 One way of preserving food is to pickle it in acid, as this kills bacteria. Suggest a reason why most bacteria cannot survive in a low pH.

⬇ Chapter summary

- Animal and plant cells have the following parts: cell membrane, cytoplasm, nucleus, mitochondria; in addition, plants cells have a cell wall, vacuole and sometimes chloroplasts.
- Cells differentiate in multicellular organisms to become specialised cells, adapted for specific functions.
- Tissues are groups of similar cells with a similar function; organs may comprise several tissues performing specific functions; organs are organised into organ systems, which work together in organisms.
- Diffusion is the passive movement of substances, down a concentration gradient.
- The cell membrane forms a selectively permeable barrier, allowing only certain substances to pass through by diffusion, most importantly oxygen and carbon dioxide.
- Visking tubing can be used as a model of a cell membrane.
- Osmosis is the diffusion of water through a selectively permeable membrane from a region of high water

(low solute) concentration to a region of low water (high solute) concentration.
- Active transport is an active process by which substances can enter cells against a concentration gradient.
- Enzymes control the chemical reactions in cells; they are proteins made by living cells, which speed up – or catalyse – the rate of chemical reactions.
- The specific shape of an enzyme enables it to function, the shape of the active site allowing it to bind to its appropriate substrate.
- Enzyme activity requires molecular collisions between the substrate and the enzyme's active site.
- Increasing temperature increases the rate of enzyme activity, up to an optimum level, after which any further increase results in the enzyme being denatured. Boiling denatures most enzymes.
- Enzyme activity varies with pH. For each enzyme, there is an optimum pH, which is different for different enzymes.

Ⓗ

► Chapter review questions

1 a) State the function of the cell membrane. [1]

b) The diagram below shows two cells that are carrying out respiration. Oxygen molecules are shown inside and outside both cells.

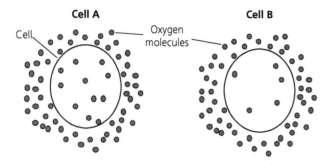

Cell A   **Cell B**

Cell   Oxygen molecules

Copy and complete the following statements by choosing the correct answer. [2]

i) In cell A:

– the oxygen molecules move into the cell

– the oxygen molecules move out of the cell

– there is no net movement.

ii) In cell B:

– the oxygen molecules move into the cell

– the oxygen molecules move out of the cell

– there is no net movement.

Now answer the following questions. [2]

iii) In which cell would there be greater net movement of oxygen? [1]

iv) Name the process by which the oxygen molecules are moving. [1]

(from WJEC Paper B2(H), Summer 2014, question 1)

2 A student used red blood cells to carry out an investigation into cell membranes. Red blood cells were placed in salt solutions at three different concentrations. A sample of red blood cells was then removed from each concentration and placed on a microscope slide. The cells were viewed using a microscope for a period of time. The observations were recorded in a table.

Concentration of salt solution, %	Observation of rd blood cells
0.0	Swell and burst
0.9	Remain the same size
3.0	Smaller and shrivelled

Explain the observations shown in the table. [6]

(from WJEC Paper B2(H), Summer 2014, question 9)

3 The graph below shows the result of an investigation into the effect of pH on the action of two digestive enzymes labelled **A** and **B**.

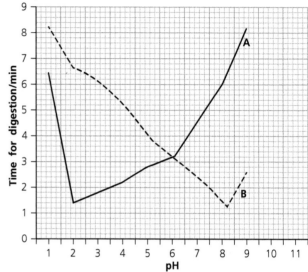

a) From the graph, state the time taken for enzyme **B** to complete its digestion at pH4.5. [1]

b) At what pH is the rate of reaction the same for both enzymes? [1]

c) From the graph, describe and explain the effect of pH on enzyme **A**. [4]

(from WJEC Paper B2(H), January 2013, question 1)

4 *Valonia ventricosa* is an unusual single-celled organism which lives in the seas of tropical and subtropical areas. It lives in shallow depths (80 m or less). The single cell is large, up to around 5 cm. The cell has a cellulose cell wall, a vacuole and many nuclei and chloroplasts. It attaches to rocks by small hair-like structures called rhizoids. Its large size makes it easy to study and scientists have measured the concentrations of ions in the vacuole and the surrounding sea water. The results for some ions are shown below.

Ion	Concentration	
	Cell vacuole	Sea water
Potassium	0.5	0.01
Calcium	0.002	0.01
Sodium	0.1	0.5

a) State three features of *Valonia* that are found in plant cells. [3]

b) State two features of *Valonia* that are different from a normal plant cell. [2]

c) Look at the concentration data. Suggest, with reasons, how each of the ions enters the cell (by diffusion or by active transport). [4]

d) What is the ratio of potassium in the cell vacuole compared to the sea water? [1]

2 Respiration and the respiratory system in humans

🏠 | **Specification coverage**

This chapter covers the GCSE Biology specification section
1.2 Respiration and the respiratory system in humans **and**
GCSE Science (Double Award) specification section
1.2 Respiration and the respiratory system in humans.

It covers the processes of aerobic and anaerobic respiration,
along with the respiratory system, which enables the oxygen
required for respiration to be taken to the tissues and the
carbon dioxide produced to be removed.

▶ Why study respiration?

Every cell in every organ of every living organism on this planet
needs energy. The energy is produced by the breakdown of food
molecules, which store chemical energy. Respiration is the process
in which the food is broken down and the energy is released for use.
If a cell stops respiring, it dies. Many other processes are needed for
life, but respiration has to continue 24 hours a day, seven days a
week, for the whole of the organism's life.

Respiration occurs in every living cell. The usual food molecule
respired is glucose (although it is possible to use others) and
oxygen is used up in the process if respiration is aerobic, which
it usually is. Carbon dioxide and water are produced as waste
materials from the process. The word equation for aerobic
respiration is:

$$\text{glucose} + \text{oxygen} \rightarrow \text{carbon dioxide} + \text{water} + \text{ENERGY}$$

The equation is a summary of the process, but it is an oversimplification. Aerobic respiration is actually a complex series of
chemical reactions, each one controlled by a different enzyme.

The chemical energy in the glucose is gradually extracted by the
reactions of respiration, and temporarily stored in a compound
called adenosine triphosphate, usually known by its shortened
name, ATP. This chemical releases its energy wherever it is required
in a cell.

How can we measure respiration rate?

We can measure respiration rate in a number of ways.

> We can measure the uptake of oxygen (which is possible but difficult).
> We can measure the production of carbon dioxide (which is easy).
> We can measure the energy given off as heat during respiration. The useful energy produced in respiration is not heat, but chemical energy. However, whenever you get an energy change, some energy is lost as heat. The quantity of heat lost will be related to the amount of respiration occurring.

This experiment measures the heat given off by germinating seeds. In a germinating seed, there is rapid growth, so a lot of respiration goes on to provide the cells with the energy they need to grow.

Apparatus

> 2 thermos flasks
> mung beans, previously soaked in water
> 2 thermometers
> cotton wool
> disinfectant

Procedure

1 Set up the two thermos flasks as shown in Figure 2.1. Flask A is the experimental flask; flask B is a control, using boiled (dead) seeds.
2 Record the temperature of each flask.
3 Leave for 24 hours.
4 Record the temperatures again.

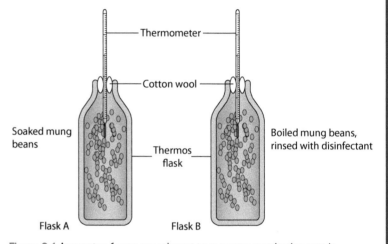

Figure 2.1 Apparatus for an experiment to measure respiration rate in germinating mung beans.

Analysing your results

1 Explain your results for flask A.
2 Explain the purpose of flask B.
3 Why were the seeds in flask B rinsed in disinfectant? (Think about what is likely to happen to dead seeds.)
4 Why were the seeds in flask A not rinsed in disinfectant?
5 Although there were roughly the same number of seeds in flask A and flask B, it is not necessary to have the same number (or the same mass) of beans in each flask. Why not?
6 Suggest a reason why it would not be a good idea to leave the seeds for longer than 24 hours before taking the second reading.

2   Respiration and the respiratory system in humans

▶ Can respiration happen without oxygen?

Cells do not always have a ready supply of oxygen. Some organisms live in places that are anaerobic (without oxygen) or where oxygen levels are very low. Even in humans and other mammals, oxygen levels in certain tissues can get very low (for example, in muscle tissue during strenuous exercise). Yet these cells survive.

They survive because they can respire **anaerobically**. Even without oxygen, certain cells can partially break down glucose and release some of the energy from it.

In anaerobic respiration in animals, glucose is broken down into lactic acid, and the word equation is simple:

$$glucose \rightarrow lactic\ acid + ENERGY$$

Anaerobic respiration is a much less efficient process than aerobic respiration, because the glucose is not fully broken down and so much less ATP is formed for each molecule of glucose used. For that reason, animal cells always respire aerobically when they can, and only use anaerobic respiration as a standby when oxygen is in short supply or unavailable (Figure 2.2).

What is oxygen debt?

If you run fast, you get breathless. When you stop, you breathe faster and deeper for a while. What you are doing is paying back your oxygen debt. During vigorous exercise, your breathing cannot supply your muscles with all the oxygen they need, and so they switch to anaerobic respiration. As a result, lactic acid builds up. This is not good, because it can cause your muscles to ache, and there is also still quite a lot of energy locked up in it (remember that the glucose is not fully broken down in anaerobic respiration). Oxygen breaks down lactic acid and releases the remaining energy. So, when you finish the exercise, your body keeps breathing faster and deeper to provide extra oxygen to break down the lactic acid. In effect, you are breathing in the oxygen that you needed (but couldn't get) during the exercise. You have built up an oxygen debt, which is then repaid after the exercise is finished.

▶ What's the difference between respiration and breathing?

People sometimes get confused between respiration and breathing, thinking that they are two words for the same thing. They are not. The fact that the organs we use to breathe are part of what we call the 'respiratory' system does not really help.

▶ **Respiration** is the process that goes on in all living cells, releasing energy from food molecules (usually glucose) to provide for the cell's energy needs. In general, oxygen is needed for this.
▶ **Breathing** is the way some animals get the oxygen they need for respiration. Plants don't breathe, and in fact many animals don't either. Many small animals can absorb oxygen through the surface of their body.

Figure 2.2 During a sprint, the athlete's muscle cells will not be able to get all the oxygen they need because of the huge demand for energy. The muscles respire anaerobically so that the sprinter's legs can keep moving.

💬 Discussion point

Evolution usually eliminates inefficient processes in favour of more efficient ones, yet anaerobic respiration is still quite common, in certain cells or in whole organisms. Why do you think that is?

Why do we need a respiratory system?

The function of the respiratory system is to extract oxygen from the air and get it into the blood, from where it can travel to all cells in the body. The respiratory system also removes carbon dioxide, which is a waste product of respiration. But why do we need a respiratory system when some animals manage without one?

Animals without a respiratory system are all quite small. They absorb oxygen through their skin (or cell membrane, in the case of single-celled animals). From the surface, the oxygen diffuses to all the cells in the animal. As these animals are small, there aren't many cells to supply with oxygen, and none of them is very far away from the surface. Diffusion is a slow process, but in such small animals it is fast enough, because of the small distance the oxygen has to travel. In larger animals, this will not work. If we absorbed oxygen through our skin, the cells deep inside our body would die before any oxygen had a chance to reach them.

In addition, as animals get larger their surface area : volume ratio reduces. As the surface area is a measure of the supply of oxygen and volume is a measure of demand, it means that the supply cannot meet the demand in larger animals.

Another factor is that larger animals tend to be more active, and so have a greater demand for oxygen, because they expend more energy.

All large animals therefore have a respiratory system of some sort, and these systems all have similar characteristics:

▶ They have a very **large surface area** for their size. Because oxygen enters through the surface of the respiratory organs, the more surface there is, the more oxygen can enter.
▶ The respiratory surface is **thin**, so that it is easy for oxygen to diffuse through it.
▶ The surface is **moist**, because oxygen needs to dissolve in order to get through into the blood. In the winter, you may have seen the result of this moisture, when you breathe out into cold air, and your breath appears as a sort of mist of condensed water droplets.
▶ The respiratory organs are **well supplied with blood vessels**, because blood is needed to carry the absorbed oxygen away to the tissues.

▶ What makes up the human respiratory system?

The respiratory system of a human is shown in Figure 2.3.

Figure 2.3 The human respiratory system.

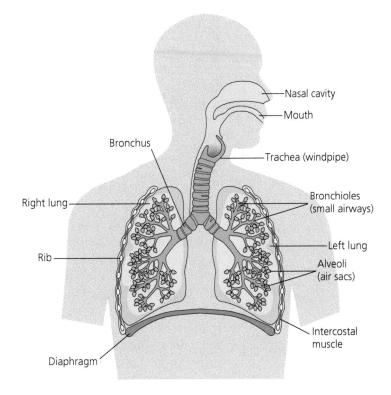

Air travels into the body when we breathe in via the nose and mouth. It enters the lungs through the trachea, which splits into two bronchi (singular: bronchus), one going to each lung. Each bronchus splits into a number of smaller tubes, the bronchioles, which eventually end in a cluster of alveoli (singular: alveolus) (Figure 2.4). The respiratory system is protected by the ribs. The lungs are inflated and deflated using muscles – the intercostal muscles and the diaphragm.

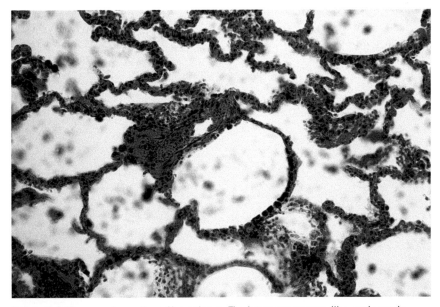

Figure 2.4 Microscopic section of lung tissue. The lungs are sponge-like, and mostly composed of air.

How do we breathe?

When the lungs expand, they suck air in; when they contract, they push air out again. There is no muscle in the lungs, though, so they cannot move on their own. The mechanism of breathing relies on the diaphragm, which is a sheet of muscle underneath the rib cage, and the rib cage itself, which is moved by the intercostal muscles between the ribs. It is also important to realise that the lungs are **elastic** (springy).

Breathing out is easiest to understand. When we breathe out, the intercostal muscles move the rib cage **downwards and inwards**, and the diaphragm moves **upwards**. This decreases the volume of the thorax and puts pressure on the lungs, so that the air in them is 'squeezed' out.

Breathing in is the reverse process. The rib cage is moved **upwards and outwards**, and the diaphragm **flattens**. This increases the volume of the thorax, and the lungs, because they are elastic, will naturally expand. The expansion of the lungs sucks air in through the trachea.

The movement of air into and out of the lungs is a result of differences in pressure between the air inside the lungs and the outside air. Gases always move from areas of higher pressure to areas of lower pressure. The breathing mechanism creates a low pressure inside the lungs (lower than the outside air) when breathing in, and a pressure that is higher than the outside air when breathing out.

The breathing mechanism is summarised in Figure 2.5. Note that, during inspiration (breathing in), both the intercostal muscles and the diaphragm are contracted, and during expiration (breathing out) all the muscles are relaxed.

Expiration is aided by the elasticity of the lungs. When they are not being stretched by air flowing in, they naturally recoil to help push air out.

Breathing in (inspiration)

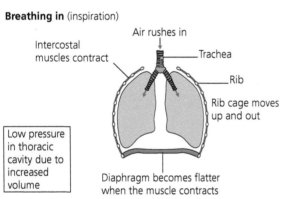

Breathing out (expiration)

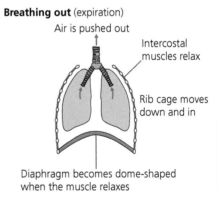

Figure 2.5 Mechanism for breathing in and out.

This mechanism can be illustrated using an artificial model of the respiratory system, as shown in Figure 2.6. The lungs are represented by the balloons, the rib cage by the bell jar, and the diaphragm by the rubber sheet. This is, in effect, two models in one. It is a model of the respiratory system as well as a model of the respiratory mechanism.

Figure 2.6 Bell jar model of the respiratory system.

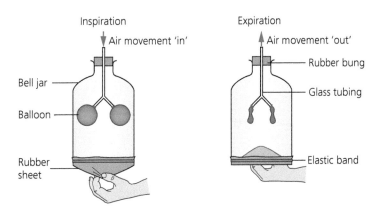

How are the lungs protected from infection?

Wherever the inside of the body is open to the air, there is always a possibility of getting infections from microbes in the atmosphere. Body openings need to have some sort of protective mechanism, and in the lungs this is provided in the trachea and the bronchi. The cells lining these tubes produce mucus, which is a sticky substance that traps dust and microbes from the air as it passes through. However, this is not enough on its own. The mucus could sink down into the lungs and the trapped dust and microbes could still cause irritation and infection. This is prevented by the cilia on the cells lining the breathing tubes. Cilia are small hair-like structures, and they constantly move, pushing the mucus up towards the top of the trachea, where it can be swallowed and (eventually) eliminated from the body via the digestive system. These cilia are shown in Figure 2.7.

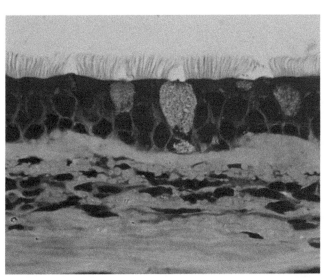
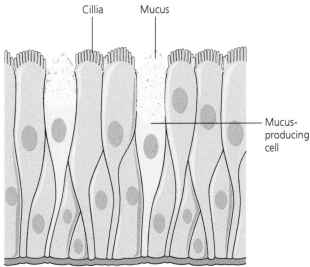

Figure 2.7 Photomicrograph and diagram showing the cells lining the trachea and bronchi.

Test yourself

5 Explain why very small organisms do not need respiratory systems.
6 Put these structures in the correct order, in terms of the way air passes through them: bronchiole, trachea, alveolus, bronchus.
7 In what direction does the diaphragm move when we breathe in?
8 Why is it important that the gas exchange surface is well supplied with blood?
9 When the lungs expand, why does this draw air into them?

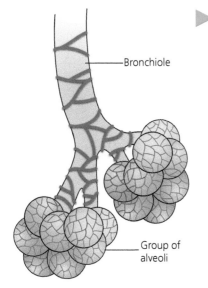

Bronchiole

Group of alveoli

Figure 2.8 A cluster of alveoli at the end of a bronchiole. The millions of tiny alveoli give the lungs a huge surface area for gas exchange.

How are gases exchanged in the lungs?

As you go deeper into the lungs, the tubes become narrower and thinner walled. The trachea and the bronchi, because they are relatively large in diameter, need support, which is provided by rings of cartilage. The smaller tubes, the bronchioles, are narrow and do not need this support. Each bronchiole ends in a group of thin-walled sacs, or alveoli (singular: alveolus). It is in the alveoli, and only there, that gases are exchanged – oxygen goes out, into the blood, and carbon dioxide goes in.

Alveoli are ideal for gas exchange. They have a huge total surface area (the surface area of an adult human's lungs is roughly the size of a tennis court), they have very thin, moist walls and they are surrounded by blood capillaries (Figure 2.8).

Gas exchange occurs through the alveolus wall by diffusion. Oxygen diffuses from the air (where it is more concentrated) into the blood (where it is less concentrated). This difference in concentrations is called a **concentration gradient**. The blood carries the oxygen away from the alveolus, and the air content of the alveolus is refreshed with each breath, so that the concentration gradient is always maintained. For carbon dioxide, the situation is reversed, and it moves from the blood into the alveolus (Figure 2.9).

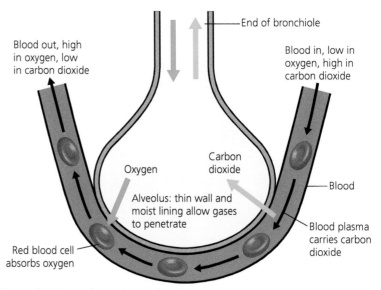

End of bronchiole

Blood out, high in oxygen, low in carbon dioxide

Blood in, low in oxygen, high in carbon dioxide

Carbon dioxide

Oxygen

Blood

Alveolus: thin wall and moist lining allow gases to penetrate

Blood plasma carries carbon dioxide

Red blood cell absorbs oxygen

Figure 2.9 Gas exchange in the alveolus

How does the air we breathe in differ from the air we breathe out?

It is not true to say that we breathe in oxygen and we breathe out carbon dioxide. We breathe in air, and we breathe out air, but the composition of that air is different. The approximate figures are given in Table 2.1.

Table 2.1 Approximate composition of inspired and expired air.

Gas	% in inspired air	% in expired air
Oxygen	21	16
Carbon dioxide	0.04	4
Nitrogen	79	79

You can see that even in expired air there is a significant amount of oxygen, but the percentage is lower than in inspired air because some has been absorbed at the alveoli, and replaced by carbon dioxide. The percentage of nitrogen remains unchanged because the body does not use that gas.

In addition, the expired air contains more water vapour than inspired air, because the surfaces of the alveoli are moist and the air absorbs some water vapour while it is in the alveoli. As the internal temperature of the body is (usually) higher than that of the surrounding air, at 37 °C, the expired air also tends to be warmer than inhaled air.

Practical

Comparing carbon dioxide in inspired and expired air

The apparatus shown in Figure 2.10 allows you to breathe in through one of the tubes (A) and out through the other (B). Limewater is placed in both tubes to test for carbon dioxide. Carbon dioxide turns limewater milky, although the test is not sensitive to small volumes of the gas.

Safety notes

Limewater (calcium hydroxide) is harmful if swallowed.

Procedure

1 Put the two pieces of rubber tubing in your mouth and breathe in and out gently through your mouth. When you breathe in, bubbles will be seen in tube A. When you breathe out, bubbles will be seen in tube B.

2 Continue to breathe in and out and observe the limewater to see when it goes cloudy.

Analysing your results

1 What are your conclusions from this experiment?

2 A student put forward a hypothesis that 'There is carbon dioxide in expired air but not in inspired air'.

   a) Explain why the evidence from this experiment cannot support this hypothesis.

   b) Suggest how you could modify the procedure to test this hypothesis.

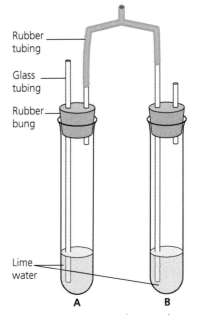

Figure 2.10 Apparatus to detect carbon dioxide in inspired and expired air.

Discussion point

Why does inspired air come in through tube A and go out through tube B?

How does smoking damage your lungs?

The worst thing you can do to your respiratory system is to smoke. When you smoke, you inhale a mixture of over 4000 chemicals from the tobacco, and many of these are harmful. They include:

▶ 43 chemicals known to cause cancer (**carcinogens**)

▶ **tar**, which is a sticky substance that clogs up the small air passages and alveoli in the lungs

▶ **nicotine**, which harms the body in a variety of ways and is a highly addictive substance – it is the nicotine that gets smokers 'hooked'

- **carbon monoxide**, a poisonous gas that makes it more difficult for the red blood cells to carry oxygen
- a variety of other harmful substances including ammonia, formaldehyde, hydrogen cyanide and arsenic.

Smoking is a known cause of diseases of the respiratory system, such as:

- **Lung cancer** – 90% of lung cancers are thought to be caused by smoking. One in ten moderate smokers and one in five heavy smokers die from the disease.
- **Emphysema** – The chemicals in tobacco smoke damage the walls of the alveoli, and eventually they break down. This means the alveoli can no longer be used to exchange gases, and the body suffers from low levels of oxygen. This causes breathing difficulties and can result in death (Figure 2.11).

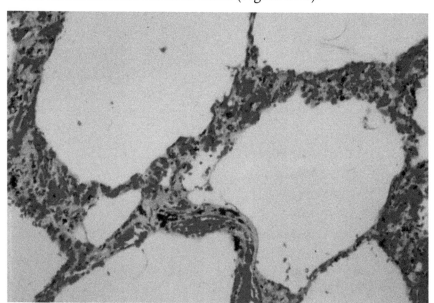

Figure 2.11 Lung tissue damaged by emphysema. The dark spots are cavities, caused by the bursting of alveoli. Their colour is due to tar deposits.

As well as causing lung disease, the chemicals in the smoke also disable the respiratory system's defence mechanisms. As we saw earlier, the trachea and bronchial tubes are lined with mucus, which keeps them moist and traps harmful particles, such as dust. In order to stop the mucus gradually sinking down into the lungs, it needs to be constantly moved upwards (to eventually be swallowed into the oesophagus) by the cilia on the surface of the cells lining the tubes (Figure 2.12). The

Figure 2.12 Cilia on the surface of the cells lining the respiratory tubes.

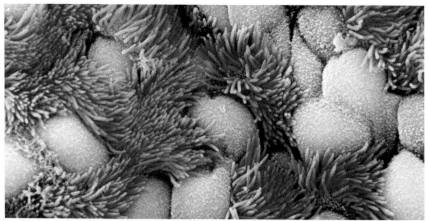

→ | Activity

The evidence linking smoking and lung cancer

Evidence has been accumulated over the last 60 years about the links between smoking and lung cancer. It is well known that lung cancer is much more common in smokers, and that smokers can reduce their risk of lung cancer by giving up smoking (Figure 2.13).

Questions

In Figure 2.13, compare the data for a lifelong non-smoker with that for someone who gave up at age 30.

1 Roughly how much more likely is the smoker who quit at 30 to develop lung cancer by the age of 75?
2 Suggest a reason why, if you smoke, it is best to give up before the age of 40.

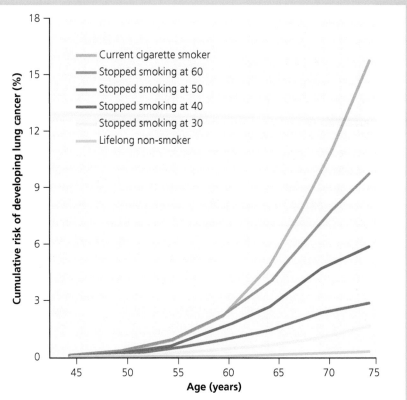

Figure 2.13 Effects of stopping smoking at various ages on the cumulative risk (%) of developing lung cancer by age 75 (for men).

cilia are paralysed by the chemicals in cigarette smoke. This paralysis means that harmful substances, including tar from the tobacco smoke, can now enter the smaller bronchioles and the alveoli. A further defence mechanism is coughing to try to 'blast' the irritants out of the lungs, and this is why smokers cough. Coughing, however, can also damage the alveoli. The cilia remain paralysed for about 20 minutes, and if cigarettes are smoked this often over a long period of time, they remain paralysed, and eventually they will die off. They will then only regenerate if the smoker quits.

Smoking is linked to a wide variety of other diseases too, including many that are not connected with the respiratory system, like heart disease, stroke and cancers of the mouth, bladder, oesophagus, kidney and pancreas.

💬 | Discussion point

Overall, smokers are about 15 times more likely to get lung cancer than non-smokers. In itself, however, this does not prove that smoking *causes* lung cancer. Why not, and what extra evidence is needed to show a causal link?

✔ | Test yourself

10 Why does no gas exchange happen in the bronchioles?
11 State two ways in which the respiratory system maintains an oxygen concentration gradient between the alveoli and the blood.
12 Why is it not true to say we breathe in oxygen and we breathe out carbon dioxide?
13 State three diseases that can be caused by smoking tobacco.
14 Explain why paralysis of the cilia caused by smoking is potentially harmful to the body.

Chapter summary

- Aerobic respiration is a series of enzyme-controlled reactions that occur in cells when oxygen is available.
- Aerobic respiration uses glucose and oxygen to release energy, and produces carbon dioxide and water.
- Energy is released in the form of ATP.
- Anaerobic respiration occurs when oxygen is not available. In animals, glucose is broken down into lactic acid.
- During strenuous exercise, anaerobic respiration in muscles builds up an oxygen debt, which is repaid after the exercise by breathing faster and deeper than normal.
- Anaerobic respiration is less efficient than aerobic respiration, and produces less ATP per molecule of glucose.
- Larger animals need a respiratory system because simple diffusion over the surface cannot supply the increased volume of the organism with oxygen, and diffusion is too slow to reach the centre of the organism.
- The human respiratory system consists of the following structures: nasal cavity, trachea, bronchi, bronchioles, alveoli, lungs, diaphragm, ribs and intercostal muscles.
- Mucus lining the respiratory system traps dust and microbes. The cilia on the cells of the breathing tubes move the mucus up to the top of the trachea, where it can be swallowed.
- The cilia are paralysed by tobacco smoke, so that the mucus sinks into the lungs, carrying the dust and microbes with it.
- Movements of the ribs and diaphragm cause breathing in (inspiration) and breathing out (expiration).
- Movement of air takes place due to differences in pressure between the lungs and the outside of the body.
- A bell jar model can be used to illustrate inspiration and expiration, but the model has limitations.
- Gas exchange occurs at the alveoli, which have thin walls, a moist lining and a good blood supply.
- Expired air contains more carbon dioxide and less oxygen than inspired air.
- Smoking is a major contributory factor in lung cancer and emphysema.

▶ Chapter review questions

1 a) When an athlete starts to run a race, aerobic respiration is taking place in her muscle cells. Write a word equation to show this process. [1]

  b) Towards the end of the race anaerobic respiration is taking place in her muscle cells. Write a word equation to show this process. [1]

  c) Explain why aerobic respiration is more efficient than anaerobic respiration. [1]

(from WJEC Paper B2(H), Summer 2011, question 7)

2 The diagram below shows the human respiratory system.

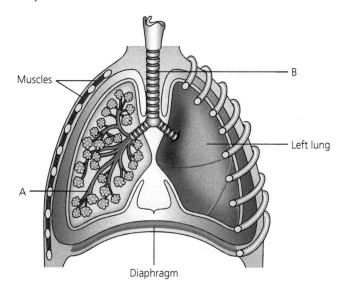

a) Name structures **A** and **B** on the diagram. [2]

b) A person's breathing rate was measured on a spirometer for 120 seconds.

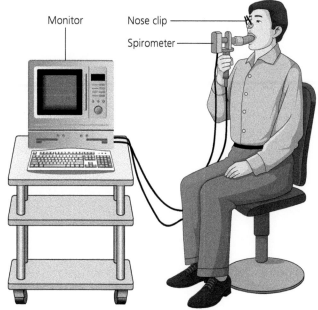

The person breathed normally, then took deep breaths and then breathed normally again. A graph of this breathing pattern was printed and is shown below.

Use the graph to:

  i) calculate the normal breathing rate for this person [1]

  ii) calculate the difference between the volume of air inspired during a single normal breath and the *second* deep breath taken. [2]

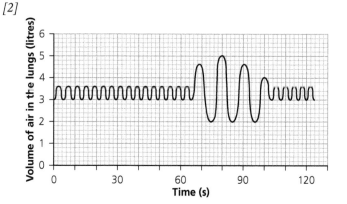

(from WJEC Paper B2(H), Summer 2015, question 3)

3 Digestion and the digestive system in humans

 Specification coverage

This chapter covers the GCSE Biology specification section 1.3 Digestion and the digestive system in humans and GCSE Science (Double Award) specification section 1.3 Digestion and the digestive system in humans.

It covers the need for digestion, the structure of the digestive system in humans and the mechanisms by which larger molecules are broken down into smaller soluble molecules that can be absorbed into the blood. There is also consideration of a balanced diet and the effects of excess sugar and fat in the diet.

▶ Why do we digest food?

Humans and all other animals get their energy from food. When we eat food, it enters the gut, which is basically a tube that goes through the body – to be of any use, that food has to move out of the gut and into the blood system, which then takes it to all parts of the body. There are two ways in which most of the food we eat needs to be changed so that it can get out of the gut and into the blood system.

1 The large molecules in the food have to be broken down into small molecules, which can be absorbed through the wall of the gut.
2 Insoluble molecules in the food have to be changed into water-soluble ones, so they can dissolve in the blood and be transported around.

The process of digestion makes these changes, breaking down complex food molecules into small, soluble ones. It is worth noting that some of the molecules we eat don't need digesting – they are already small and soluble (for example, glucose and vitamins), but these are the exceptions rather than the rule.

▶ What foods need digesting?

The complex food molecules in our diet fall into three categories:

▶ **fats**, which are broken down into glycerol and fatty acids
▶ **proteins**, which are broken down into amino acids
▶ some **carbohydrates**, the main one being starch, which is broken down into the simple sugar, glucose.

Glycerol, fatty acids, amino acids and glucose can all be absorbed readily into the blood. Not all these end products are used for energy, however. Energy is provided by glucose, glycerol and fatty acids, but amino acids are not normally respired. Instead they are used as raw materials for making new proteins, which are needed for growth.

Practical

How do we know what our food contains?

There are chemical tests for a number of the different food groups, including proteins and carbohydrates (with specific tests for starch and for glucose).

> **Test for protein** — A small volume of dilute sodium hydroxide solution is added to the test solution, then a roughly equal volume of copper sulfate solution is added. If protein is present, a purple colour is seen. This is called the Biuret test. Sometimes, the sodium hydroxide and the copper sulfate are combined into one solution, known as Biuret solution.

> **Test for starch** — When iodine in potassium iodide solution is added to starch, the brown colour of the iodine turns to blue-black.

> **Test for glucose** — When a solution containing glucose is heated with blue Benedict's solution, a reddish-orange precipitate is formed. This is called the Benedict's test. The more glucose there is, the more precipitate is formed. As more and more precipitate is formed, the blue colour turns first to green, then orange, then to brick red. Some tests are described as qualitative, because they only tell you whether a substance is present or not. Some tests are quantitative and give you a measurement of how much of a substance is present. This test is called semi-quantitative, because it gives an idea (but not a precise measurement) of how much glucose is present.

Investigating the Benedict's test for sugar

The full instructions for the Benedict's test are given here. It would be useful if you had done, or seen, the Benedict's test before attempting the questions which follow.

Apparatus

> 250 cm^3 beaker
> boiling tube
> tripod
> gauze
> Bunsen burner
> test-tube holder
> heating mat
> test solution
> Benedict's solution

Safety notes

Wear eye protection.

Procedure

1 Place some test solution into the boiling tube. Make sure the tube is under half full.
2 Add enough Benedict's solution to give a distinct blue colour.
3 Half fill the beaker with water and bring it to the boil using the Bunsen burner (Figure 3.1). Alternatively, you can use a water bath at a temperature of 80 °C.
4 Using the test-tube holder, place the boiling tube in the beaker.
5 Boil for 5 minutes, and observe the colour changes.

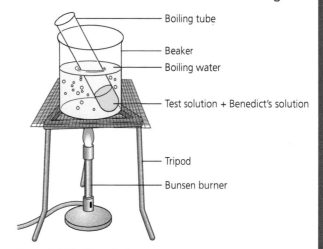

Figure 3.1 The Benedict's test.

Questions

1 The more glucose there is in the solution, the further the colour change in the Benedict's solution goes. This gives an indication, but not a measure, of how much glucose Is In the solution. Suggest how you might get an actual measure of the concentration of glucose in a solution, using the Benedict's test.
2 How accurate do you think your measure (from question 1) would be? Give reasons for your answer.

▶ Where in the body is food digested?

Food is digested in the digestive system, which is basically a tube called the **gut** that goes through the body. The useful products are absorbed into the blood as the food moves through the digestive system, and eventually the non-digestible remains are egested at

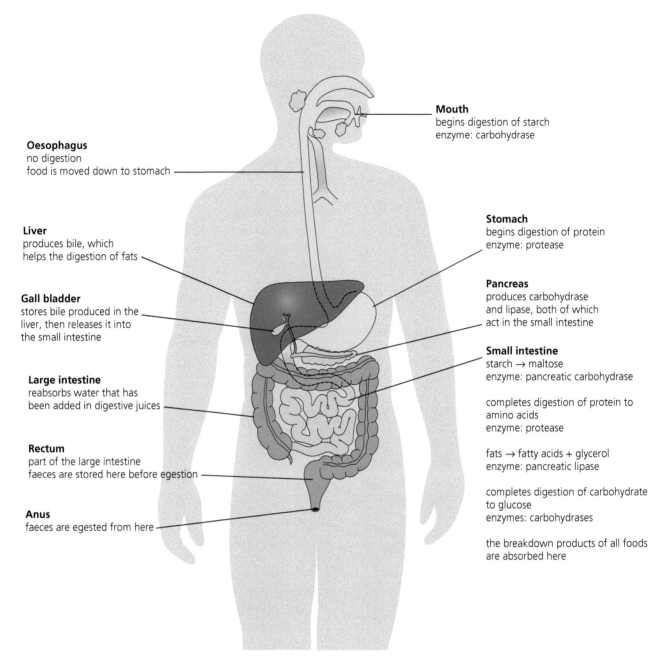

Mouth
begins digestion of starch
enzyme: carbohydrase

Oesophagus
no digestion
food is moved down to stomach

Stomach
begins digestion of protein
enzyme: protease

Liver
produces bile, which
helps the digestion of fats

Pancreas
produces carbohydrase
and lipase, both of which
act in the small intestine

Gall bladder
stores bile produced in the
liver, then releases it into
the small intestine

Small intestine
starch → maltose
enzyme: pancreatic carbohydrase

completes digestion of protein to
amino acids
enzyme: protease

Large intestine
reabsorbs water that has
been added in digestive juices

fats → fatty acids + glycerol
enzyme: pancreatic lipase

Rectum
part of the large intestine
faeces are stored here before egestion

completes digestion of carbohydrate
to glucose
enzymes: carbohydrases

Anus
faeces are egested from here

the breakdown products of all foods
are absorbed here

Figure 3.2 The human digestive system.

the other end of the gut. Different parts of the gut are specialised for specific functions. Figure 3.2 shows the functions of the various parts of the digestive system. As well as the gut, the digestive system also includes some associated organs (the liver, gall bladder and pancreas). There are three processes, which occur in different parts of the digestive system:

1 **digestion** – mainly in the mouth, stomach and small intestine
2 **absorption** – mainly in the small intestine (food) and large intestine (water)
3 **egestion** – in the rectum and anus.

Bile is a digestive juice that is produced by the liver and stored in the **gall bladder**, from where it is released when necessary. We shall see what it does later in this chapter.

By the time the food reaches the second half of the small intestine, digestion is complete and the breakdown products are all small, soluble molecules. These can then be absorbed into the blood. To help this, the walls of the small intestine are covered in small, finger-like projections called villi, which greatly increase the surface area over which food can be absorbed (Figure 3.3).

By the time the gut contents reach the large intestine, all of the useful products have been absorbed into the blood. What remains is indigestible material, which will eventually be egested from the body as faeces. However, as it enters the large intestine, the material is a liquid, because of the digestive juices that have been added as it travelled down the gut. If this liquid was eliminated with the faeces, the body would soon dehydrate. The job of the large intestine is to reabsorb water from the waste, which therefore solidifies as it goes down the large intestine. The now-solid faeces are temporarily stored in the rectum before being egested via the anus.

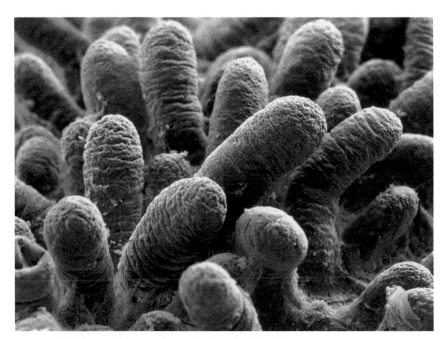

Figure 3.3 Surface of the small intestine, showing the villi.

✔ Test yourself

1 Why does the body not need to digest vitamins?
2 Is the iodine test for starch qualitative or quantitative?
3 Name the chemicals that are formed by the digestion of proteins.
4 Where in the digestive system is protein digested?
5 Which part of the digestive system reabsorbs water that has been added to the gut in the digestive juices?

How is food moved through your gut?

In order to push food along your digestive system, waves of muscle contraction constantly move along the gut. These waves are called peristalsis (Figure 3.4).

When the circular muscles contract just behind the food, this squeezes the food forward, rather like squeezing toothpaste from a tube.

Figure 3.4 Peristalsis in the gut.

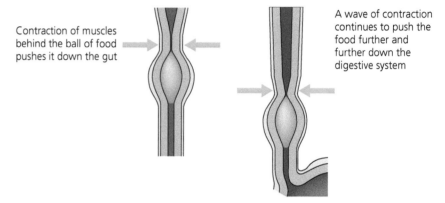

Contraction of muscles behind the ball of food pushes it down the gut

A wave of contraction continues to push the food further and further down the digestive system

Peristalsis happens throughout the gut, including in your stomach. It is only here that you are sometimes aware of the process. The stomach has rings of muscle that contract to close the openings at the top and bottom, so food inside cannot move out of the stomach. Peristalsis churns the food around, mixing it with digestive juices containing enzymes, which helps with digestion. So when the stomach is full, peristalsis simply moves the food around, and this makes no noise. If your stomach is empty of food, however, the only thing that can be squeezed by peristalsis is the air inside. Moving air around in the stomach creates a sort of gurgling sound – the 'rumbling stomach' that we associate with being hungry.

What does bile do?

Bile is produced by the liver and stored in the gall bladder. When a meal containing fat is being digested, the gall bladder releases bile down the **bile duct** into the small intestine. Bile is not an enzyme, but its job is to help the lipase enzyme in the small intestine to digest fats. Bile **emulsifies** fat, splitting it into small droplets and allowing a greater surface area for the lipase enzyme to work on (Figure 3.5).

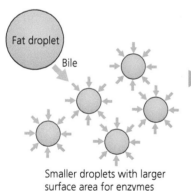

Fat droplet

Bile

Smaller droplets with larger surface area for enzymes

Figure 3.5 The effect of bile on fats.

How do nutrients get out of the gut and into your body?

Once food substances have been digested into small, soluble chemicals, they can get through the wall of the gut and into your bloodstream, which will take them all around the body.

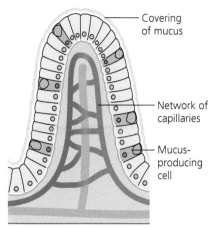

- Covering of mucus
- Network of capillaries
- Mucus-producing cell

This happens in the small intestine, and is helped by the fact that the small intestine is lined with projections called villi, described earlier (Figure 3.3). The villi vastly increase the surface area over which food can be absorbed. Remember that food is constantly moving through the intestine, so it is important that it is absorbed fairly quickly (even though the small intestine is long – about 7 m long in an adult). The structure of a villus is shown in Figure 3.6.

Figure 3.6 The structure of a villus.

Practical

What are the optimum conditions for lipase enzymes?

In this experiment, liquid detergent is added to simulate the activity of bile. Like bile, detergent is an emulsifier. If available, bile salts could be substituted for the liquid detergent.

Safety notes

Make sure you do not get any of the phenolphthalein indicator on your skin or in your eyes.

Apparatus

> full-fat milk or single cream
> phenolphthalein indicator
> 5% lipase solution
> sodium carbonate solution, 0.05 mol/dm^3
> liquid detergent
> water baths set at 30 °C, 40 °C, 50 °C and 60 °C
> ice
> test-tube rack
> 2 × 10 cm^3 measuring cylinders
> 2 × 100 cm^3 beakers
> 2 × 250 cm^3 beakers
> 6 thermometers
> 12 test tubes
> glass rod
> 2 cm^3 syringe
> stopclock or stopwatch

Procedure

Water baths must be set up in advance. Electronic water baths can be used for the temperatures above room temperature. Use the 250 cm^3 beakers to make water baths at 10 °C (cold water + ice) and 20 °C (tepid water). A beaker of lipase solution should be placed in each of the water baths, so that lipase is available at each of the test temperatures.

1 Add 5 drops of phenolphthalein indicator to two test tubes, for the first temperature.
2 Add 5 cm^3 of milk to each of the two test tubes.
3 Add 7 cm^3 of sodium carbonate solution to the test tubes, which should turn the phenolphthalein pink. The sodium carbonate is added to produce alkaline conditions, which are best for lipase.
4 Add 1 drop of liquid detergent to one of the test tubes.
5 Place both test tubes in the appropriate water bath. Test the temperature with the thermometer and leave until the contents of the tubes have reached the experimental temperature.
6 Add 1 cm^3 of lipase to each tube and start the stopwatch.
7 Stir the tubes gently, looking for the indicator to lose its pink colour.
8 Record the time taken for the colour change (in seconds).
9 Repeat steps 1 to 8 for each of the other temperatures.
10 Record all your results in a table.
11 Plot a graph of temperature against time taken for the colour change. The graph should have two lines on it, one for the enzyme with liquid detergent and one for the enzyme without detergent.

Analysing your results

1 What effect did temperature have on the enzyme activity?
2 What effect did the liquid detergent have on enzyme activity?
3 Looking at the results, do you think they were accurate? Give a reason for your answer.
4 Suggest one possible source of inaccuracy in this experiment.

Test yourself

6 What name is given to the process which moves food along the gut?
7 What is the function of the lacteals in the villi of the small intestine?
8 How does bile help the digestion of fats?
9 What is the advantage to the body of having villi on the wall of the small intestine?

▶ What is a 'balanced diet'?

Humans require a variety of nutrients, as each type has a slightly different function in the body.

Practical

Using Visking tubing as a model gut

Visking tubing can be used to model the gut. This is because it will allow small molecules to pass through it and prevent the passage of larger ones, just like the lining of the gut does. The model has limitations, though. Visking tubing is non-living and lets small molecules through because it has microscopic pores in it. The gut lining does not have pores and the nutrients have to pass though cell membranes and cytoplasm in order to get to the capillaries. The transport of molecules through the gut wall is therefore much more complex than transport through Visking tubing.

Apparatus

> boiling tube
> length of Visking tubing, knotted at one end
> dropping pipette
> elastic band
> iodine in potassium iodide solution
> 1% starch solution
> 1% glucose solution
> test-tube rack
> Benedict's solution or Clinistix (if using Benedict's solution, you will need apparatus to carry out the Benedict's test, as described earlier in this chapter, in the practical 'Investigating the Benedict's test for sugar')

Safety notes

Wear eye protection.

Procedure

1 Set up the apparatus as shown in Figure 3.7. The volumes of the starch and glucose solutions in the Visking tubing should be roughly equal, but the exact volumes used are unimportant. Leave the apparatus for 30 minutes.
2 Test the liquid in the boiling tube for starch by adding iodine in potassium iodide solution. A blue-black colour indicates the presence of starch.
3 Test the liquid in the boiling tube for glucose using the Benedict's test or Clinistix (which change colour if glucose is present).

Question

Explain how this experiment models the action of the small intestine.

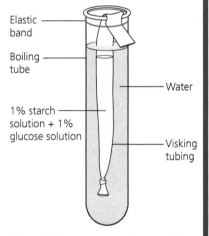

Figure 3.7 Apparatus used to model absorption in the gut.

Labels: Elastic band; Boiling tube; Water; 1% starch solution + 1% glucose solution; Visking tubing

▸ **Glucose**, formed by the breakdown of carbohydrates, is the main energy provider in the body, being broken down by respiration in the cells.

▸ **Fatty acids and glycerol** from fats also provide energy. Fats actually contain more energy per gram than glucose, but it can only be released slowly. For this reason, fats are useful as an energy store.

▸ **Amino acids** from proteins are re-assembled in the body into new proteins, to form many useful products or to be used for making new cells in growth.

Apart from the chemicals that are formed by the digestion of food, there are other useful substances in the diet that can be absorbed directly because they are small molecules.

▸ **Minerals** have a variety of functions – for example, iron is needed to make haemoglobin, the blood pigment that carries oxygen around the body.

▸ **Vitamins** also perform various jobs, and are often needed for important chemical reactions to take place. Vitamin C, for example, has many benefits, including helping the immune system to function properly, and contributing to the health of teeth and gums.

▸ **Water** is important because it is the main constituent of cells, and all of the chemical reactions in the body involve chemicals that must be dissolved in water. In addition to drinking water, humans get water from eating, as most foods contain quite a lot of it.

Finally, there are health benefits from having quite a lot of **fibre** in the diet. Fibre is indigestible carbohydrate, so it never enters the bloodstream. However, it provides bulk for the gut to act on during peristalsis, and so aids the efficient movement of food through the gut.

Balancing the diet

The National Health Service has published what are called 'reference intakes' – that is, the recommended amount of each food type that should be consumed each day. These are based on an average-sized woman doing an average amount of exercise, and are as follows:

▸ **energy** – 8400 kJ or 2000 kcal
▸ **fat** – 70 g (with no more than 20 g of saturated fat)
▸ **carbohydrates** – 260 g, with no more that 90 g in the form of sugar
▸ **protein** – 50 g
▸ **salt** – 6 g.

Figure 3.8 In order to have a healthy diet, all of the nutrients you need should be in the correct balance. Although all are beneficial, it is possible to have too much of certain food groups – for example, carbohydrates, fats and certain minerals.

How can diet affect your health?

In order to survive and keep active, we need energy. Energy is needed for all living processes, and the more physical activity we do, the more energy we need. Different foods contain different amounts of energy, but the main type of food the body uses for energy is **glucose**, a sugar that we get from eating any carbohydrate. If we eat more carbohydrate than we need at the time, the body stores it in the liver for future use as a substance called glycogen. If we keep eating more food than we need, this store becomes full. The body then changes the carbohydrate into **fat**, which is stored under the skin and around the internal organs. In other words, we 'get fat'. These fat stores also increase directly if excess fat is eaten. If we eat less and exercise more, these stores get used up, but, because the body uses the glycogen store before the fat, it can take some time.

If someone is overweight, it means that they have been taking in more energy than they need, probably for some time. To lose weight it is usually necessary to cut down on the energy taken in and also take more exercise to increase the energy used. To keep the weight down, a new, healthier food and exercise regime will have to be continued throughout life.

Being overweight or obese leads to an increased likelihood of heart disease, stroke, cancer and type 2 diabetes.

A problem is that many processed foods that you can buy contain high levels of sugar and fat (including saturated fats, which are particularly bad for you). Sugar is used as a preservative and so you often find it added to foods where you might not expect it – for example, in sauces, ready meals and crisps. Other foods – such as chocolate, biscuits, cakes, dairy-based desserts and fizzy drinks – are more obviously high in sugar.

High fat foods include cheese and other dairy products, meat, fried food, crisps, biscuits and cakes. Vegetable oils, although high in calories, are less harmful than animal fats as they do not contribute to the build-up of fat on the walls of the blood vessels, which can lead to heart attacks and strokes.

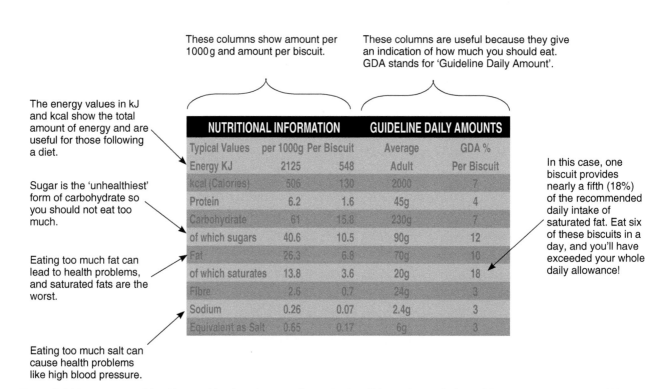

These columns show amount per 1000g and amount per biscuit.

These columns are useful because they give an indication of how much you should eat. GDA stands for 'Guideline Daily Amount'.

The energy values in kJ and kcal show the total amount of energy and are useful for those following a diet.

Sugar is the 'unhealthiest' form of carbohydrate so you should not eat too much.

Eating too much fat can lead to health problems, and saturated fats are the worst.

Eating too much salt can cause health problems like high blood pressure.

In this case, one biscuit provides nearly a fifth (18%) of the recommended daily intake of saturated fat. Eat six of these biscuits in a day, and you'll have exceeded your whole daily allowance!

NUTRITIONAL INFORMATION			GUIDELINE DAILY AMOUNTS	
Typical Values	per 1000g	Per Biscuit	Average Adult	GDA % Per Biscuit
Energy KJ	2125	548		
kcal (Calories)	506	130	2000	7
Protein	6.2	1.6	45g	4
Carbohydrate	61	15.8	230g	7
of which sugars	40.6	10.5	90g	12
Fat	26.3	6.8	70g	10
of which saturates	13.8	3.6	20g	18
Fibre	2.6	0.7	24g	3
Sodium	0.26	0.07	2.4g	3
Equivalent as Salt	0.65	0.17	6g	3

Figure 3.9 Most processed food has a table of nutrients on the packaging. This can be useful in deciding how much of it should be included in the diet.

Food additives are another cause for concern. The main problem is salt, which is added to very many foods, both for taste and as a preservative. Excess salt in the diet can lead to high blood pressure and increase the risk of heart disease and strokes.

How much energy does food contain?

People can tell how much energy is in packaged food by looking at a nutrient table, but how is this figure arrived at? Food scientists get the figures by burning the food so that the energy is released as heat. They measure the heat given off to see how much energy a known mass of food produces.

A special piece of apparatus called a food calorimeter (or bomb calorimeter) is used for this. Figure 3.10 shows an example.

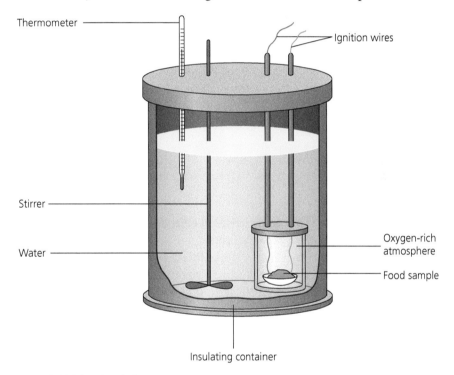

Thermometer

Ignition wires

Stirrer

Water

Oxygen-rich atmosphere

Food sample

Insulating container

Figure 3.10 A bomb calorimeter.

It is very important that the food is burnt completely (to get all the energy from it) so the sample is burnt in an oxygen-rich atmosphere. The heat given off is measured by the increase in temperature of the water. The stirrer ensures that the heat is evenly spread and the container is insulated to minimise heat loss to the atmosphere.

Of the foods groups in our diet, fat has the highest energy content. Carbohydrates and proteins have a similar energy content, but our bodies do not generally use protein as an energy source. There is no way protein can be stored in the body, so the only source of protein as an energy source would be to break down our own cells, which would not be a good idea (although it can happen in starvation).

Investigation into the energy content of foods

When food is burnt, the chemical energy in the food is released as heat. Measuring the heat given off tells us how much energy the food contains.

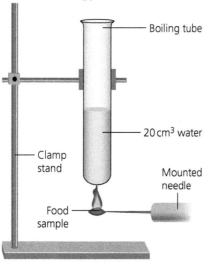

Figure 3.11 Apparatus to investigate the energy content of foods.

Apparatus

> samples of food – for example, crisps or other potato snacks, biscuits, pasta, breakfast cereal
> (Any food used should be packaged, so that the energy content obtained experimentally can be compared with the value given on the labelling. Nuts should not be used due to the possibility of allergic reactions.)
> boiling tube
> clamp stand, boss and clamp
> Bunsen burner
> heatproof mat
> measuring cylinder, 50 cm³
> mounted needle with wooden handle
> tongs or forceps
> thermometer

Safety notes

Wear eye protection.

Procedure

1 Set up the apparatus as shown in Figure 3.11.
2 Measure 20 cm³ of water into the boiling tube.
3 Measure the temperature of the water with the thermometer and record it.
4 Weigh a sample of food and record the mass.
5 Impale the food sample carefully on a mounted needle. If this is not possible, pick it up using the tongs or forceps.
6 Hold the food in the Bunsen burner flame until it catches alight.

7 As soon as the food is alight, hold it under the boiling tube of water.
8 Hold the food in place until it has burnt completely. If the flame goes out, try to light it again. If it will not light, you can assume that all the possible energy has been extracted.
9 As soon as the food has burnt completely, measure and record the temperature of the water again.
10 Repeat the procedure at least twice more for the same food, replacing the water each time.
11 Calculate the rise in temperature each time.
12 Calculate the energy released from each food sample by using this formula.

$$\text{energy released from food, per gram (J)} = \frac{\text{mass of water (g)} \times \text{temperature rise (°C)} \times 4.2}{\text{mass of food sample (g)}}$$

4.2 is the specific heat capacity of water, in joules per gram per °C – that is, the number of joules required to raise the temperature of 1 g of water by 1 °C. 1 cm³ of water has a mass of 1 g.

13 Calculate a mean value for the energy content of the food, in joules per gram.
14 Compare the mean value you obtain with the energy content per gram from the food packaging. (It will probably be given as energy per 100 g on the packet.)
15 If there is time, repeat the procedure using a different food.
16 Compare your results with those of other groups to get a better idea of repeatability of the technique. Here you are assessing repeatability, not reproducibility (they may have used other foods), so you just need to look at how similar the repeat results were, not the actual figures.

Questions

The values obtained from this technique are usually inaccurate. These are possible sources of error compared with using the calorimeter:

> The food is burnt in air rather than oxygen, which may mean that the food is not completely burnt.
> The water is not stirred.
> The apparatus is not insulated.
> The thermometer bulb is very near to the heat.
> A lot of heat escapes into the surrounding air.
1 How serious an error do you think each of the above factors may have caused in your experiment?
2 How could the apparatus be modified to reduce some of these errors?
3 How could you discover how inaccurate the calculated value might be?

Test yourself

10 Why can eating too much carbohydrate make you overweight?

11 Why are proteins particularly important in the diet of young children?

12 Why is sugar often added to meat-based ready meals?

13 What problems can arise if your diet is too high in salt?

14 What units are used to measure the energy content of foods?

 Chapter summary

- Complex, insoluble food molecules need to be broken down into small, soluble molecules that can enter the blood system.
- This breakdown, in the digestive system, is called digestion.
- Digestion is aided by enzymes.
- Visking tubing behaves similarly to the wall of the gut, and it can be used as a 'model gut'.
- Fats are digested into fatty acids and glycerol.
- Proteins are digested into amino acids.
- Starch is digested into glucose.
- The test for starch uses iodine in potassium iodide solution, which turns blue-black.
- The test for glucose is the Benedict's test. The test sample is boiled with Benedict's solution (blue) and, if glucose is present, a reddish-orange precipitate is formed.
- The test for protein is called the Biuret test. Copper sulfate and sodium hydroxide are added to the test solution. If protein is present, a purple colour appears.
- The digestive system consists of the mouth, oesophagus, stomach, small intestine, large intestine, anus, liver, gall bladder and pancreas.
- The mouth contains carbohydrase, which digests starch.
- The stomach contains protease, which digests proteins.
- The small intestine contains a variety of enzymes, which complete the digestion of carbohydrates, proteins and fats.

- The liver produces bile, which is stored in, and released from, the gall bladder.
- Bile emulsifies fats, which aids their digestion.
- Food is moved along the digestive system by peristalsis.
- Glucose from carbohydrates, and fatty acids and glycerol from fats, provide energy for the body.
- Amino acids from proteins form the building blocks for new proteins, which are needed for growth and repair of tissues and organs.
- For optimum health, we need to eat a balanced diet, with appropriate levels of carbohydrates, fats, proteins, minerals, vitamins, water and fibre.
- Carbohydrates, fats and proteins all contain energy. Fats contain the most, with carbohydrates and proteins having less (and roughly equal) amounts.
- Our bodies use carbohydrates and fats for energy. If we take in too much fat or carbohydrate, the extra energy is stored as fat.
- A diet that is too high in sugar or fat can lead to health problems.
- A number of different additives are added to our food. Salt is a common one, and eating too much salt can lead to high blood pressure and associated health problems.

► Chapter review questions

1 The diagram shows the digestive system.

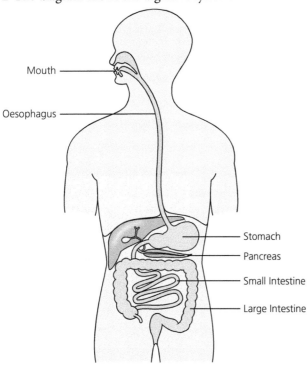

Mouth

Oesophagus

Stomach

Pancreas

Small Intestine

Large Intestine

a) Write down the name of the organ from the diagram that best fits each description below.

   i) An organ that secretes lipase, proteases and carbohydrases. [1]

   ii) The organ where fats are digested to fatty acids and glycerol. [1]

   iii) The organ where digestion of starch begins. [1]

b) The diagram below represents a short length of a starch molecule.

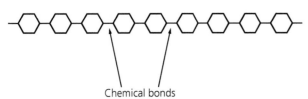

Chemical bonds

   i) Name the type of enzyme that digests the chemical bonds in a starch molecule. [1]

   ii) Name the end product of starch digestion. [1]

(from WJEC Paper B2(H), January 2012, question 1)

2 The following statement refers to a process that occurs in the digestive system.

   'The muscles in front of the food relax whilst the muscles behind the food contract.'

a) Name the process being described. [1]

b) The graph shows the results of an investigation into the activity of an enzyme at various pH levels. The enzyme was acting on a food substance and the mass of this food substance remaining undigested at each of the pH levels was recorded.

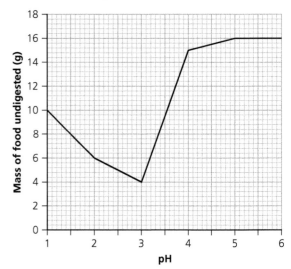

   i) State what happens to the mass of undigested food from pH3 to pH6. [2]

   ii) State the optimum pH of this enzyme. [1]

   iii) Name the organ in the human body where this enzyme is found and name the class of food it acts on. [2]

(from WJEC Paper B2(H), Summer 2015, question 5)

3 John is a severely obese 27-year-old man. He weighs 31 stone and takes no exercise. For his height, John should weigh about 14 stone. A typical lunch for John would include:

- 2 double cheeseburgers
- 2 litre bottle of cola.

The table below shows the nutrition facts for one double cheeseburger and one litre of cola. It also shows the Guideline Daily Amounts (GDA) of the various nutrients for an adult man.

	Guideline Daily Amount (GDA)	Double cheeseburger, per 220 g serving	Cola, per litre
Energy, in kcal	2500.0	1120.0	400.0
Carbohydrate, in g	300.0	47.0	108.0
of which sugars, in g	70.0	8.0	108.0
Fat, in g	95.0	105.6	0.0
Protein, in g	55.0	25.0	0.4
Sodium, from salt, in g	2.4	2.0	0.12

Using the information and data above, and your own knowledge, describe the ways in which John's lifestyle and diet could lead to health problems. [6]

(from WJEC Paper B1(H), Summer 2015, question 4)

The circulatory system in humans

 Specification coverage

This chapter covers the GCSE Biology specification section
1.4 Circulatory system in human and GCSE Science (Double
Award) specification section 1.4 Circulatory system in humans.

It covers the structure and function of the circulatory system
and blood in humans. The differences between the different
types of blood vessel are covered, along with an evaluation of
the different treatments for cardiovascular disease.

▶ Why do we have a circulatory system?

Humans and other mammals are complex organisms, with many
specialised organs in different parts of their bodies. These different
organs need to be coordinated, and must therefore be able to
communicate with each other. One way of communicating is by
using chemical messengers (hormones), which are made in certain
organs and transported to others where they cause specific effects.
The body needs to transport other substances from one place to
another too. For example, food molecules absorbed from the gut
and oxygen entering the body in the lungs must be transported to
every respiring cell. Waste products need to be carried to the lungs
(in the case of carbon dioxide) and the kidneys (in the case of urea) in
order to be removed by excretion. All these substances move around
the body in the circulatory system, carried in the blood through a
complex network of blood vessels that reaches every part of the body.

Transport is an important function of the blood, but because the
blood system reaches every part of the body it is also able to play a
useful role in the immune system, which protects the body from
infection. Pathogens cannot locate themselves anywhere that is
very far from the blood system. Immunity is therefore a second
important function of the circulatory system.

The blood is moved around the circulatory system by the heart,
which is a pump made of muscle that contracts repeatedly to push
the blood through the blood vessels.

> **Key term**
>
> Pathogen A disease-causing organism
> (of any type).

▶ What is in blood?

Blood is a liquid, which allows it to be moved easily around the
body. It also contains cells, and both the cells and the liquid part
of the blood play vital roles in its function. The components of the
blood are as follows.

▶ **Plasma** – the liquid part of the blood, which transports water-
soluble substances including digested food, carbon dioxide, urea,
salts and hormones.

- ▶ **Red blood cells** – these cells carry oxygen around the body, attached to the red pigment haemoglobin.
- ▶ **White blood cells** – there are various types of white blood cells, including phagocytes, which engulf and destroy bacteria. Other white blood cells make antibodies, which destroy pathogens in other ways. The white blood cells are mainly responsible for the immunity function of the blood.
- ▶ **Platelets** – these are cell fragments that help the blood to clot. Clotting plugs wounds and so helps to prevent infection.

The appearance of these structures under the microscope is shown in Figure 4.1.

Figure 4.1 A blood 'smear' as seen under a light microscope.

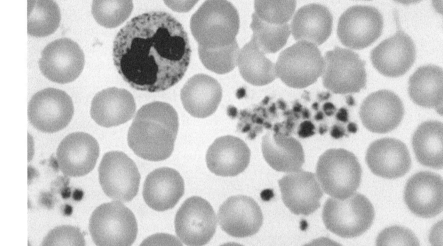

Side

Surface

Figure 4.2 Structure of a red blood cell.

Figure 4.2 shows the structure of red blood cells. They are biconcave discs – round and flattened, with a central indentation. Their flattened, biconcave shape increases the surface area for absorption of oxygen, compared with a rounded shape. They are red because they contain the red blood pigment haemoglobin, which absorbs oxygen. Mature red blood cells are unusual because they have lost their nucleus, which allows more haemoglobin to be packed into their cytoplasm.

The structure of a phagocyte is shown in Figure 4.3. These cells ingest bacteria.

White blood cells can change shape and can also move. This allows them to squeeze through tiny gaps in capillary walls and enter the tissue fluid to fight infection in tissues.

Engulfed bacterium

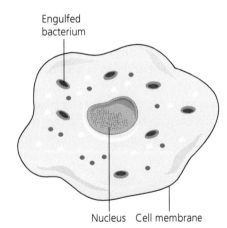

Nucleus   Cell membrane

Figure 4.3 Structure of a phagocyte (a type of white blood cell).

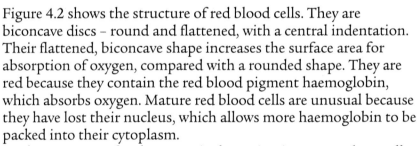

✔ Test yourself

1 What are the two main functions of the blood?
2 What is the function of the platelets?
3 Which substance is transported in the blood, but not in the plasma?
4 State two ways in which red blood cells are adapted for their function.
5 Some people have too few red blood cells (a condition called anaemia). One symptom is a constant feeling of tiredness. Suggest a reason for this.

▶ How does blood circulate around the body?

The blood is pumped by the heart, and moves around the body in blood vessels called arteries, veins and capillaries. When it leaves the heart the blood travels to the organs in arteries. The arteries branch into a large number of small capillaries, which take the blood through the organs. The capillaries then join to form veins, which carry the blood back to the heart. The structure of the circulatory system of a mammal is shown in Figure 4.4. Note that the diagram has been simplified. The aorta actually branches into several different arteries, each of which goes to a particular organ. The blood is then transported out of the organs by different veins, which join to form the vena cava.

Figure 4.4 Structure of the circulatory system of a mammal. The arrows show the direction of blood flow.

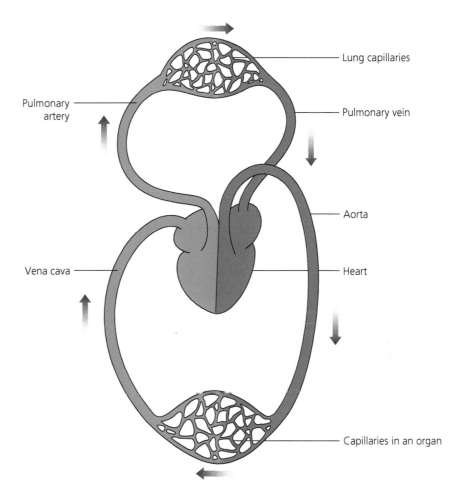

Lung capillaries

Pulmonary artery

Pulmonary vein

Aorta

Vena cava

Heart

Capillaries in an organ

You can see that in mammals the heart is divided into two halves. The left half receives blood from the lungs and pumps it to the rest of the body. The right half receives blood from the body and pumps it to the lungs. The blood therefore travels in two separate circuits around the body (in the **pulmonary circulation** to and from the lungs, and the **systemic circulation** to and from the rest of the body), and for that reason the circulatory system is referred to as a **double circulation**. In any one complete circuit of the body, the blood travels through the heart twice.

▶ How does the heart work?

Blood is moved around the body by the pumping of the heart. The heart can act as a pump because it is made of muscle, which, when it contracts, applies a force to the blood and pushes it out into the arteries. The muscle in the heart is a special type, which is not found anywhere else in the body. It can contract on its own without any nerve stimulation, and it does not get tired, which is just as well because if you live to be about 80 your heart will have beaten non-stop approximately 3 billion times!

The exterior of the heart is shown in Figure 4.5. You can see that the outside of the heart has its own blood supply, via the coronary artery. Even though the heart is filled with blood, the muscular walls are so thick that the outside needs a separate blood supply. The blood supplies the nutrients and oxygen that the heart needs to keep beating.

Figure 4.5 Exterior view of the human heart.

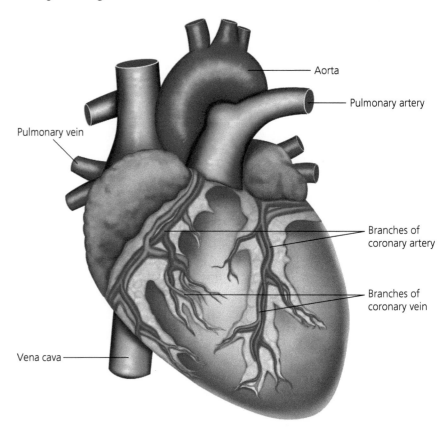

Aorta

Pulmonary artery

Pulmonary vein

Branches of coronary artery

Branches of coronary vein

Vena cava

4 The circulatory system in humans

Key term

De-oxygenated blood Blood that has had most of its oxygen removed. The opposite of oxygenated blood, which contains a lot of oxygen.

Figure 4.6 shows the way in which blood flows through the heart. The heart is, in effect, two pumps side-by-side. The right side deals with de-oxygenated blood and the left side with oxygenated blood. The mechanisms on the two sides of the heart are the same, although the names of the structures are different. Another difference is that the left ventricle has a much thicker wall than the right ventricle, because it has to pump blood all around the body, whereas the right ventricle only has to pump blood to the lungs (which are very close to the heart). Notice that the right and left sides of the heart refer to the right and left of the person whose heart it is, not the viewer's right and left.

Figure 4.6 Blood flow through the heart.

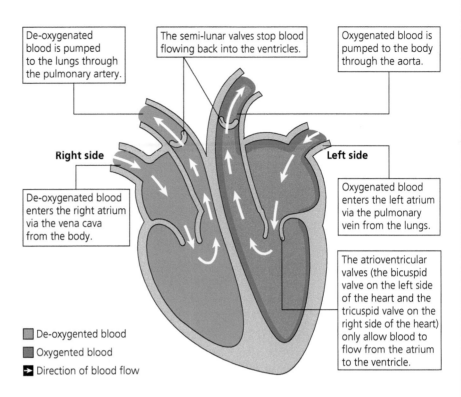

De-oxygenated blood is pumped to the lungs through the pulmonary artery.

The semi-lunar valves stop blood flowing back into the ventricles.

Oxygenated blood is pumped to the body through the aorta.

Right side

Left side

De-oxygenated blood enters the right atrium via the vena cava from the body.

Oxygenated blood enters the left atrium via the pulmonary vein from the lungs.

The atrioventricular valves (the bicuspid valve on the left side of the heart and the tricuspid valve on the right side of the heart) only allow blood to flow from the atrium to the ventricle.

■ De-oxygented blood
■ Oxygented blood
 Direction of blood flow

Blood has to flow through the heart in one direction only, from the atria to the ventricles and then out of the arteries at the top. Ensuring one-way flow is the job of the valves. The atrioventricular (bicuspid and tricuspid) valves between the atria and the ventricles stop backflow from the ventricles into the atria, and the semilunar valves at the beginning of the aorta and the pulmonary artery make sure that blood that has left the heart is not sucked back when the heart relaxes.

Let's follow the flow of blood through the left side of the heart:

1 Oxygenated blood flows into the left atrium via the pulmonary vein from the lungs.
2 The left atrium contracts, forcing the bicuspid valve open so that blood goes into the left ventricle.
3 The left ventricle contracts, which forces the bicuspid valve shut but opens the semilunar valve.
4 Blood flows out of the heart via the aorta.

✔ Test yourself

6 Starting from the heart, in what order does the blood flow through capillaries, arteries and veins?
7 What is a double circulation?
8 What is the name of the blood vessel that supplies blood to the outside of the heart?
9 What is the function of the semilunar valves in the heart?
10 The blood pressure in the left ventricle reaches much higher levels than in the right ventricle. Suggest a reason for this.

▶ What is the difference between arteries, veins and capillaries?

There are three main types of blood vessels – arteries, veins and capillaries. Each has a different structure, which is linked to its functions.

Arteries carry blood away from the heart. The blood is at high pressure, because the beating of the heart puts pressure on the blood. The arteries have to be able to resist that pressure.

The arteries deliver blood to the capillaries, which take blood through each of the organs of the body. The capillaries are very small and there are a very large number in each organ. It is in the capillaries that exchange of materials occurs. Oxygen and nutrients are delivered to the cells and waste products (including carbon dioxide) are picked up. Capillaries are very narrow and so the blood flows slowly through them. This, and the fact that there are so many of them, means that a lot of materials can be exchanged.

The capillaries eventually discharge their blood into veins, which take it back to the heart. By the time the blood gets to the veins, there is no longer any pulse, and the blood pressure has dropped considerably. It is important that blood is returned to the heart at the same rate as it leaves, yet veins have no pulse. Blood is moved in the veins by the muscles between which the veins run. The contraction of these muscles, while performing their functions, squeezes the thin-walled veins and so moves the blood, but not in any particular direction. The direction is controlled by the valves, which prevent the flow of blood back towards the capillaries. Arteries do not need valves because the pulse ensures that the blood flows in the right direction.

Table 4.1 Structure and function of blood vessels.

Vessel	Structural feature	Link to function
Artery	Thick muscular wall	Resists high blood pressure
	Pulse	Pushes blood through the vessel
Vein	Thinner wall than an artery	Does not need to resist high blood pressure – allows muscles around the vessel to squeeze the blood and cause it to move
	Valves	Ensures that movement of blood is only towards the heart
	Large lumen (gap in the middle of the vessel)	Increases the rate of flow of the blood
Capillary	Wall is only one cell thick	Allows easy passage of materials in and out by diffusion
	Blood flow is very slow	Allows time for materials to be exchanged
	Extensive networks in each organ	Every cell is near to a capillary – more materials can be exchanged

Table 4.1 shows the features of the different blood vessels, and how their structure links to their functions. The structure of the vessels is also shown in Figure 4.7.

Figure 4.7 Structure of an artery, a capillary and a vein (drawings are not to scale).

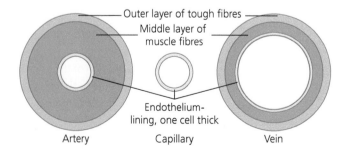

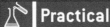 **Practical**

Observing valves in veins

Veins have valves, which only allow the blood to flow towards the heart. This can be seen in someone who has prominent veins. Males tend to have more prominent veins than females, and adults have more prominent veins than teenagers. In a young person, the most likely places to find prominent veins are on the back of the hand or the top of the foot.

Procedure

1 Place two fingers on a prominent vein on the back of the hand or the top of the foot and press.
2 Move away the finger that is furthest from the person's heart, while keeping the other finger in position, all the time maintaining pressure (Figure 4.8). The vein will flatten.

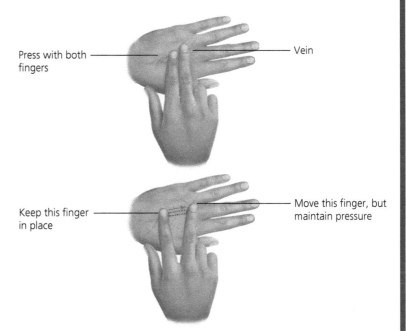

Figure 4.8 Observing valves.

3 Lift the finger that you moved. The vein should re-fill with blood.
4 Reverse the procedure, this time moving the finger nearer to the person's heart.
5 When the moved finger is lifted, the vein will only partially fill with blood. The place where the blood reaches is the position of a valve in the vein.

The procedure is illustrated in Figure 4.8.

Question

Explain what you have just seen in terms of blood flow and the action of valves.

▶ What causes cardiovascular disease?

Cardiovascular disease (CVD) is a common cause of death. CVD includes all the diseases of the heart and the circulatory system, including coronary heart disease, heart attacks, angina and strokes. CVD is usually linked to a process called atherosclerosis (Figure 4.9). This is the build-up of a substance called plaque in the walls of the arteries. Atherosclerosis has a number of effects. It makes it more difficult for blood to flow through the arteries, which means that the heart has to work harder to move the blood around. It can actually

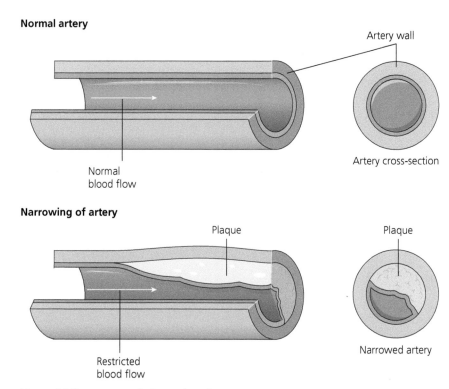

Normal artery

Artery wall

Artery cross-section

Normal blood flow

Narrowing of artery

Plaque

Plaque

Restricted blood flow

Narrowed artery

Figure 4.9 The process of atherosclerosis.

block smaller arteries, depriving the tissues they supply of oxygen and nutrients. If this happens in the coronary artery, a heart attack can result. Even if the blockage does not cause a heart attack, it often means that the person will get pains in the chest, a condition known as angina. The slower flow of blood also makes it more likely that a clot will form, which can also block off blood vessels and cause a heart attack. If the blockage happens in the brain, a stroke can result.

How can we reduce the risk of cardiovascular disease?

A number factors that increase a person's risk of developing CVD are well known, and are listed below. Some you can avoid, but others you cannot.

▸ **High blood pressure** – If you have high blood pressure it is an indication that your heart is having to work harder than is ideal, which puts a strain on both your heart and your blood vessels. One cause of high blood pressure is a diet that is too high in salt.

- **Smoking** – Tobacco smoke contains carbon monoxide, which limits how much oxygen the blood can absorb. To get the right amount of oxygen to the tissues, the heart has to work harder. Smoking can also lead to high blood cholesterol (see below).
- **High blood cholesterol** – One form of cholesterol is the substance that forms plaques in the artery walls, so a high-cholesterol diet can increase the likelihood of atherosclerosis. A high-cholesterol diet is one that contains too much saturated fat (which the body converts into cholesterol).
- **Diabetes** – Both type 1 and type 2 diabetes increase the risk of CVD.
- **Being overweight or obese** – Excess weight is linked with a build-up of fat around the organs including the heart, and the heart has to work harder to provide the energy to move the extra bulk around.
- **Lack of exercise** – Exercise improves the condition of the heart and also helps to avoid becoming overweight or obese.
- **A family history of heart disease** – Close relatives dying of heart disease increase the risk of CVD, suggesting that a person's genes can make heart disease more or less likely.
- **Ethnic background** – People of a South Asian or African Caribbean ethnicity have a statistically higher risk of CVD than people from other ethnic backgrounds.

Obviously, individuals cannot change their family history or their ethnicity, and type 1 diabetes is difficult to avoid. People with those issues can significantly reduce their risk of CVD, however, by ensuring they avoid the other factors.

How is cardiovascular disease treated?

People diagnosed with cardiovascular disease can have various treatments to relieve their condition. Some of those treatments are listed below.

- **Changes to lifestyle related to diet and exercise** – People with CVD are advised to exercise regularly and to adopt a healthy diet, so reducing their risk factors. This is particularly effective if the patient is overweight or if these factors were a contributor to the development of their condition. The dangers of surgery and drug side effects are avoided, but the new regime requires maintained willpower on the part of the patient.
- **Taking statins** – A group of drugs called statins are very effective at lowering blood cholesterol and so reducing the risk of cardiovascular disease. They are taken in the form of tablets and so the treatment is very easy. They have a very good safety record, but taking any drug can have side effects. In the case of statins, these have nearly always been both rare and mild (for example, headaches), and the British Heart Foundation states that the risk of potentially dangerous side effects is low (1 in 10 000 patients). Statins do not prevent or cure cardiovascular disease, but simply reduce one of the major risk factors.
- **Angioplasty** – Narrowed arteries can be widened by carrying out a coronary angioplasty. In this procedure, a short wire-mesh tube, called a stent, is inserted into the problem artery. A small balloon is attached to the front of the stent and is inflated to

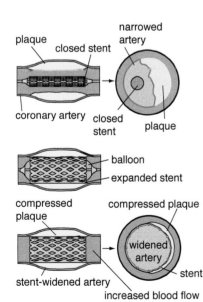

Figure 4.10 This stent allows blood to flow freely again.

widen the artery. The stent expands and holds the widened artery open. The balloon is then deflated and removed. The stent is left permanently in place. The surgery is relatively minor, although any surgery carries some risk. The operation has a good long-term success record, but if the patient does not address other risk factors then atherosclerosis may re-occur.

✔ Test yourself

11 Why do arteries have a much thicker wall than veins?
12 Why do veins have valves, but arteries do not?
13 Why is a high level of blood cholesterol a risk factor for cardiovascular disease?
14 Suggest two ways in which a patient with cardiovascular disease may be asked to alter or check their diet. Give reasons for your answers.

⬇ Chapter summary

- Blood has two functions in the body – transport and immunity.
- Blood consists of plasma, red cells, white cells and platelets.
- The red cells are responsible for the transport of oxygen, attached to the red pigment haemoglobin.
- The plasma transports nutrients, hormones, carbon dioxide, salts and urea.
- The white cells combat infections. One type, called phagocytes, engulf and destroy bacteria.
- The platelets help the blood to clot.
- The heart pumps blood around the body, due to the contraction of its muscular walls.
- The coronary artery supplies the muscle of the heart with blood.
- The mammalian circulatory system is a double circulation, where blood travels through the heart twice on each circuit of the body.
- The blood leaves the heart in arteries, flows through capillaries in the organs, then back to the heart in veins.

- Blood travels through the heart by entering the atria, passing through to the ventricles, and then being pumped out into the aorta or pulmonary artery.
- The heart has two halves. The right half deals with de-oxygenated blood, the left half with oxygenated blood.
- The main blood vessels entering and leaving the heart are the pulmonary artery, the aorta, the pulmonary vein and the vena cava.
- Valves in the heart ensure the one-way flow of blood through the heart, in the right direction.
- Materials enter and leave the blood through the thin-walled capillaries.
- Arteries, capillaries and veins are all adapted in a variety of ways to perform their function.
- Certain factors related to lifestyle and genetics can affect the risk of cardiovascular disease.
- Cardiovascular disease can be treated by changes to diet and exercise, the use of statin drugs, or by surgery (angioplasty).

► Chapter review questions

1 a) Copy and complete the table below about the different parts of the blood. [4]

Part	Function
Red cell	
	Transports glucose
Phagocyte	
	Aids blood clotting

b) Explain why the centre of a red blood cell appears paler than the surrounding cytoplasm when seen through a light microscope. [2]

(from WJEC Paper B3(H), Summer 2014, question 2)

2 The diagram shows the human circulatory system.

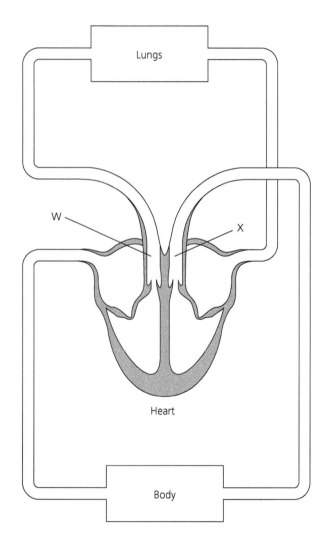

a) i) Name the blood vessels labelled **W** and **X**. [2]

ii) State two differences between the blood in vessels **W** and **X**. [2]

b) Using the diagram only, state whether the ventricles are contracting or relaxing. Explain your answer. [2]

(from WJEC Paper B3(H), Summer 2011, question 4)

3 Varicose veins are a condition which is caused by a weakening and stretching of the walls of the veins. This weakens the valves in the vein and they become less effective. The veins become swollen and enlarged, and appear blue or dark purple. This tends to be most noticeable in the legs. Symptoms include swollen feet and ankles, and aching, heavy and uncomfortable legs. The symptoms tend to be worse in warm weather or after the person has been standing for a long period of time.

a) What is the function of the valves in veins? [1]

b) Why is there no need for valves in arteries? [1]

c) Suggest why varicose veins are worse after a long period of standing. [2]

d) Suggest why the legs of a person with varicose veins feel heavy. [1]

e) Exercise is one aspect of the treatment of varicose veins. Select the most likely reason for this from the suggestions below. [1]

i) The muscles working will squeeze the veins and help to move the blood up to the heart.

ii) Exercise improves your health in many ways.

iii) The heat generated in the muscles will cause the blood to move more quickly.

iv) It means that you are not spending a long time standing still.

5 Plants and photosynthesis

 Specification coverage

This chapter covers the GCSE Biology specification section **1.5 Plants and photosynthesis** and GCSE Science (Double Award) specification section **1.5 Plants and photosynthesis.**

It covers the process of photosynthesis and factors that affect the rate of photosynthesis. Transport systems in plants are covered, along with transpiration and factors that affect transpiration.

▶ Why study photosynthesis?

Although it only happens in plants, all life on Earth depends upon photosynthesis. It is the process that converts light energy reaching the planet into food, both for plants and for the animals that form the food chains leading from those plants. It also produces oxygen as a waste product, allowing our atmosphere to support aerobic life. Scientists try to understand as much as possible about the process of photosynthesis, in the hope of being able to boost food production for the world's growing population.

▶ What do plants need to survive?

In order to carry out photosynthesis, and their other living processes, plants need certain materials from their environment. Figure 5.1 summarises their needs.

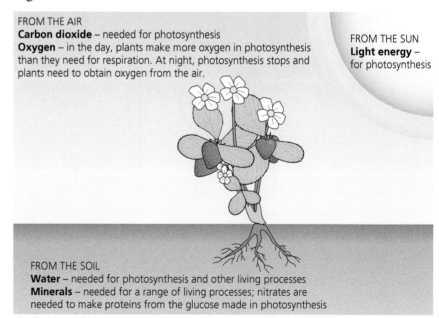

FROM THE AIR
Carbon dioxide – needed for photosynthesis
Oxygen – in the day, plants make more oxygen in photosynthesis than they need for respiration. At night, photosynthesis stops and plants need to obtain oxygen from the air.

FROM THE SUN
Light energy – for photosynthesis

FROM THE SOIL
Water – needed for photosynthesis and other living processes
Minerals – needed for a range of living processes; nitrates are needed to make proteins from the glucose made in photosynthesis

Figure 5.1 The needs of plants.

▶ How does photosynthesis work?

Photosynthesis is a complex series of chemical reactions in the chloroplasts of plant cells, but it can be summarised by the following word equation:

$$\text{carbon dioxide} + \text{water} \rightarrow \text{glucose} + \text{oxygen}$$

For the process to work, four things are needed:

- ▶ **Carbon dioxide** – Glucose is made of carbon, hydrogen and oxygen. The carbon dioxide provides the carbon and oxygen.
- ▶ **Water** – This provides the hydrogen needed to make glucose. The oxygen from the water molecules is not needed and is given off as a waste product.
- ▶ **Light** – This provides the energy for the chemical reactions in photosynthesis.
- ▶ **Chlorophyll** – The green pigment in chloroplasts is chlorophyll, which absorbs the light to provide the energy for photosynthesis.

All of the chemical reactions involved in photosynthesis are controlled by enzymes, which are available in the chloroplasts of the photosynthesising cells.

 Specified practical

Investigation into factors affecting photosynthesis

For the experiments that follow, you will test whether photosynthesis has occurred in a plant by testing its leaves for starch. Once glucose is made in a leaf, it may be used, transported to other parts of the plant, or stored as starch. For that reason, it is better to look for starch in a leaf, rather than glucose, when testing for photosynthesis.

Starch stains blue black with iodine in potassium iodide solution, but the green colours in a leaf can make it more difficult to see the stain, so first you have to remove the green chlorophyll using boiling ethanol.

Testing a leaf for starch

Apparatus

- > leaf
- > boiling tube
- > 250 cm^3 beaker
- > forceps
- > white tile
- > ethanol
- > iodine in potassium iodide solution

Safety notes

Wear eye protection.

Procedure

1. Half fill a 250 cm^3 beaker with water from a recently boiled kettle.
2. Using the forceps, dip the leaf into the boiling water for a few seconds. This kills the leaf and makes it permeable to the chemicals used later.
3. Place the leaf in the boiling tube and cover it with ethanol.
4. Place the boiling tube in the beaker of hot water and leave it for 5 minutes (Figure 5.2). The ethanol should boil, and the leaf will gradually lose colour, turning the ethanol green.

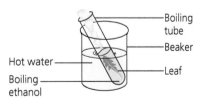

Figure 5.2 Removing chlorophyll from a leaf.

6 Using a test-tube holder, remove the boiling tube from the water bath and pour off the ethanol.

7 Remove the leaf from the boiling tube. An easy way to do this is to fill the tube with water, so that the leaf floats to the top.

8 Spread the leaf on the tile and cover it with iodine in potassium iodide solution. Leave for about a minute.

9 Gently rinse off the iodine in potassium iodide solution. Areas containing starch will now be stained blue-black.

We will now use this technique to investigate the various factors needed for photosynthesis. When doing these experiments, it is important to make sure that any starch detected has been made during the experiment, and was not already there before. To do this, the plants used (with the exception of the variegated plant used for the chlorophyll experiment) are kept in the dark for 48 hours prior to the experiment. In the dark, no photosynthesis can occur, so the plant is forced to use its stored starch for food.

Experiment 1 – Light

A leaf on a de-starched plant is partially covered with aluminium foil to prevent light reaching its surface (Figure 5.3). The plant is then left in the light for at least 24 hours, and the leaf is then tested for starch.

The covered part of the leaf remains brown. Only the exposed part turns blue-black, showing that it contains starch.

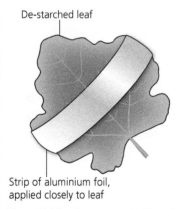

De-starched leaf

Strip of aluminium foil, applied closely to leaf

Figure 5.3 Leaf treatment for Experiment 1.

Experiment 2 – Chlorophyll

A variegated (green and white) geranium leaf, from a plant that has been kept in a well-lit place, is tested for starch. Starch is only present in the green areas, which contain chlorophyll (Figure 5.4).

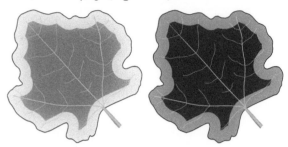

Figure 5.4 Leaf before and after testing in Experiment 2.

Experiment 3 – Carbon dioxide

A de-starched plant is set up in the light for 48 hours, as shown in Figure 5.5. The sodium hydroxide solution absorbs carbon dioxide, so that leaf A is provided with carbon dioxide, but leaf B is not.

After 48 hours, both leaves are tested for starch. Leaf A contains starch, but leaf B does not.

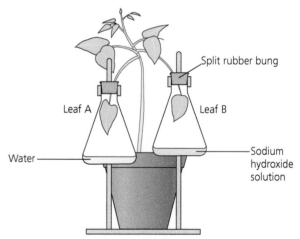

Split rubber bung

Leaf A

Leaf B

Water

Sodium hydroxide solution

Figure 5.5 Apparatus to investigate if carbon dioxide is needed for photosynthesis.

Questions

1 Why was there no need to de-starch the leaves used for Experiment 2?

2 In Experiment 3, why was leaf A put in a flask containing water?

▶ What affects the rate of photosynthesis?

Photosynthesis makes food. The more photosynthesis there is occurring in a plant, the more food it makes. Commercial plant growers obviously want photosynthesis to happen as fast as possible in their plants, because that will mean their plants will grow quicker, or be bigger or healthier. If plants are grown in greenhouses, the environmental conditions can be controlled so as to maximise photosynthesis. We know that the external factors needed for photosynthesis are light, carbon dioxide, water and a suitable temperature. So, it would seem logical that the more a plant has of these, the more photosynthesis will occur. However, things are a little more complicated than that.

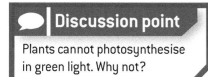

Discussion point

Plants cannot photosynthesise in green light. Why not?

Light

It is true that increasing light intensity boosts the rate of photosynthesis, but only up to a point. What no-one can alter is the amount of chlorophyll in a plant. If the light intensity is greater than the chlorophyll can absorb, then any further increase in intensity won't have any effect.

Carbon dioxide

As with light, increasing carbon dioxide levels will increase the rate of photosynthesis up to a certain level, and then increasing it further has no effect. The same argument applies as with light – once the chloroplasts have all the carbon dioxide they need, there is no benefit in increasing it.

Water

Despite the fact that water is needed for photosynthesis, increasing the amount of water available to a plant does *not* increase the rate of photosynthesis. Water is needed for much more than photosynthesis in plants, and if there is enough water to keep the plant alive, that will be enough for photosynthesis. Too much water can actually kill plants, as it drives necessary oxygen out of the soil and the roots die off.

Temperature

The chemical reactions in photosynthesis are all controlled by enzymes, and the effect of temperature on the rate of photosynthesis is due to the effect of temperature on those enzymes. Raising the temperature up to about 40°C is beneficial, as long as you don't dehydrate the plant in the process. As the temperature gets higher, though, it destroys (denatures) the enzymes and photosynthesis stops.

Limiting factors

In any set of circumstances, one factor is more important than the others in setting the rate of photosynthesis. This factor is known as the limiting factor. In different conditions, any of the factors listed above – light, carbon dioxide or temperature – can be the limiting factor. You can tell if a factor is limiting by increasing it. If the rate of photosynthesis also increases, then the factor was limiting.

What is the effect of increasing light intensity on the rate of photosynthesis?

An experiment was done using leaf discs – small circles cut from the blade of a leaf. They were placed in a syringe containing sodium bicarbonate solution (which provides the carbon dioxide needed for photosynthesis). When photosynthesis occurs, oxygen forms inside the leaf discs – this increases their buoyancy and the discs float to the surface. The apparatus was placed in a dark environment and a lamp was placed at different distances from it (Figure 5.6). The time taken for 50% of the discs to float to the surface was recorded for each distance. This time was used as a measure of the rate of photosynthesis.

The results obtained were as shown in Table 5.1 and Figure 5.7.

Table 5.1 Results of an experiment to test the effect of increasing light intensity on the rate of photosynthesis.

Distance of lamp, in cm	Time taken for 50% of discs to float, in seconds			
	Trial 1	Trial 2	Trial 3	Mean
40	848	1029	748	875.0
35	737	788	794	773.0
30	602	628	640	623.3
25	594	588	521	567.7
20	580	530	544	551.3

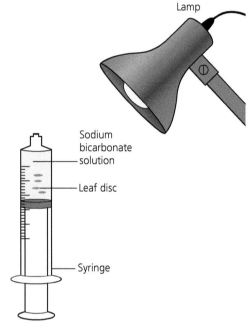

Figure 5.6 Leaf disc apparatus.

Questions

1 Why do you think the time taken for 50% of the leaf discs to float was measured in seconds rather than in minutes and seconds?
2 It was felt that measuring the time taken for 50% of the leaf discs to float would provide a more accurate measure of photosynthesis than waiting for all the discs to float. Why do you think this is?
3 Do you think the variation in the results is acceptable to draw a conclusion from? Explain your answer.
4 Do you think three repeats for this experiment is enough? Explain your answer.
5 Do you think that the differences in the results for the different distances are significant? Explain your answer.
6 What conclusion would you draw from these results?

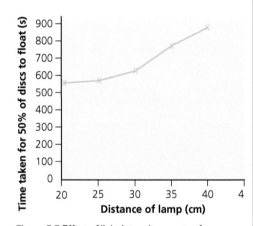

Figure 5.7 Effect of light intensity on rate of photosynthesis – graph of results.

▶ What happens to the glucose made by photosynthesis?

Just like animals, plants need a balanced 'diet', with a variety of nutrients. The difference is that they have to make the nutrients themselves (apart from minerals, which are absorbed from the soil). They need a variety of carbohydrates and proteins. They have less need for lipids, although some seeds do use oils as a food store. Carbohydrates and lipids can be made from glucose, because they contain the same chemical elements (carbon, hydrogen and oxygen). Proteins need nitrogen as well, but that is absorbed from the soil in the form of nitrates.

The main ways in which glucose is used in plants after it is formed in leaves are shown in Figure 5.8.

Figure 5.8 The fate of glucose made in photosynthesis.

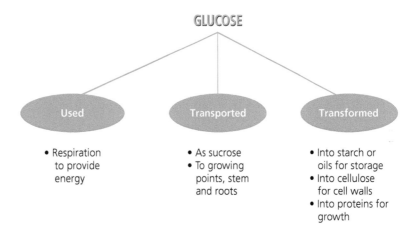

✔ Test yourself

1 Name the four factors needed for photosynthesis.
2 A student said that plants photosynthesise during the day, but respire at night. Why is this statement incorrect?
3 If you increase the light intensity shining on a plant, what will happen to the rate of photosynthesis?
4 A horticulturalist increased the level of carbon dioxide in her greenhouse by installing a burner outside and piping the carbon dioxide produced into the greenhouse. The yield of plants in the greenhouse increased. What conclusions would you draw from this?
5 Photosynthesis alone cannot supply the plant with protein. What else is necessary?

▶ What is inside a leaf?

A leaf is a complex organ, with features that make it well adapted to its function of carrying out photosynthesis. Light is absorbed by green chlorophyll, stored in chloroplasts in the leaf cells, and the structure of the leaf ensures that the cells containing the chloroplasts receive the water and carbon dioxide they need for photosynthesis. The internal structure of a leaf is shown in Figure 5.9 on the next page.

Figure 5.9 The internal structure of a leaf.

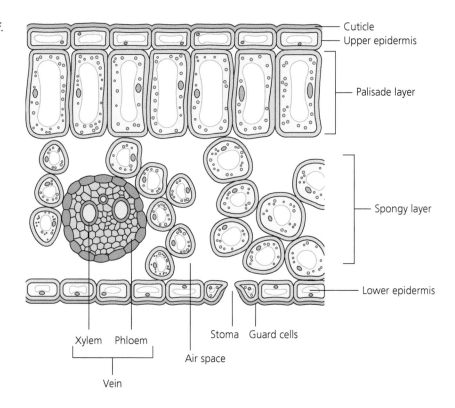

Cuticle
Upper epidermis
Palisade layer
Spongy layer
Lower epidermis

Xylem Phloem

Stoma Guard cells

Air space

Vein

The functions of each of the structures in relation to photosynthesis are given in Table 5.2.

Discussion point

Generally, plants have most or all of their stomata on the lower epidermis, away from direct sunlight so that drying out is less likely. Floating leaves, such as those of the water lily, have their stomata on the upper surface. Suggest reasons for this.

Table 5.2 Structure and function of a leaf.

Structure	Function
Cuticle	Waxy, waterproof layer that reduces water loss – it is transparent, allowing light through to the lower layers of cells, which contain chloroplasts
Palisade layer	The cells are packed with chloroplasts for photosynthesis
Spongy layer	Contains large air spaces, allowing carbon dioxide to reach palisade layer for photosynthesis, but the cells here also contain chloroplasts for photosynthesis
Vein	Contains xylem (brings water to leaf) and phloem (transports sugar away)
Guard cells	Open and close the stomata, allowing carbon dioxide to enter or preventing water loss

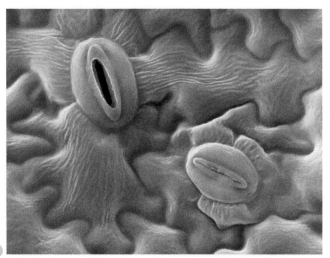

In order to let carbon dioxide in for photosynthesis, the leaf has pores called stomata (singular: stoma) that are open to the atmosphere. It is impossible to let carbon dioxide in without also allowing water to escape, and water is a valuable resource. In the daytime, water loss is bound to happen, but at night, when no photosynthesis can occur, the loss of water would be a complete waste. To reduce this water loss, the guard cells around each stoma can change shape and cause the stomata to close (Figure 5.10).

Figure 5.10 Micrographs of open (left) and closed (right) stomata.

▶ Why is water so important for plants?

In the last section, water was described as a valuable resource for plants. It is needed for photosynthesis, but there are other reasons why water is important.

▸ Cells are mostly water, and the chemicals in them need to be dissolved in water in order to carry out the reactions that make up life.
▸ All the cells in the plant need minerals, which enter via the roots. Water provides the medium that transports the minerals up the plant. It also transports dissolved sugar away from the leaves.
▸ In non-woody plants, water helps to support the plant. If the cells are turgid – that is, full of water – they are rigid and keep the plant upright. If plants don't get enough water, they will start to collapse, a phenomenon known as **wilting** (Figure 5.11).

Figure 5.11 Wilting – the effect of a lack of water.

▶ How do plants absorb water and minerals?

The water and minerals that plants need are absorbed through the surface of their roots. In living things, surfaces that absorb materials usually have features to increase their surface area, and roots are no exception. Absorption takes place in the area just behind the tip of each root, and this area is covered in root hairs (Figure 5.12). Each 'hair' is a projection of the cell wall of an epidermal cell.

We have seen in Chapter 1 that water travels into and out of cells by osmosis. Figure 5.13 (on the next page) shows how this works in a root. Water entering the root hair cell dilutes the cell sap, so that it is more dilute than that of the cell next to it. This concentration gradient means that water moves into the next cell by osmosis. The process continues, taking the water across the root.

The transport of water works because the soil water is almost always more dilute than the cell sap. However, when it comes to absorbing minerals, they may be at a low concentration in the soil water, and yet the root still needs to absorb them – that is, the minerals must move against a concentration gradient. Minerals are therefore absorbed by active transport, which uses energy to pump materials from a lower concentration to a higher concentration.

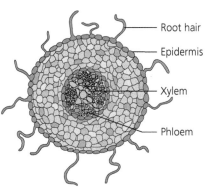

Root hair
Epidermis
Xylem
Phloem

Figure 5.12 Root hairs not only increase the surface area available for absorption, they also penetrate a little way into the soil.

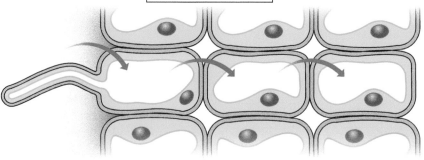

Soil water is more dilute than root hair cell sap. Water moves in by osmosis.

Root hair cell sap is diluted by water and so becomes more dilute than next cell. Water moves into next cell by osmosis.

Process is repeated and water moves further into root by osmosis.

✔ Test yourself

6 Why is it useful to the plant to have air spaces in the spongy layer of the leaf?

7 Why is it important that the stomata are closed at night?

8 Why does a house plant wilt if you don't water it?

9 Water moves across the root by osmosis because there is a concentration gradient across the root. Explain how this concentration gradient causes water movement.

10 Why would it be impossible for minerals to be absorbed by osmosis, even if the concentration outside the root was greater than inside?

▶ How does water get up the stem?

Water (together with dissolved minerals) moves up the stem in specialised tissue called xylem. Xylem cells are dead cells forming continuous tubes, which carry water from the roots up the stem and into the leaves (Figure 5.14). The xylem tubes are in the centre of the root, but then spread out towards the outer surface of the stem and give off branches that go into the veins in the leaves.

Once water gets into the xylem in the root, it is pulled up the stem by the effects of a process called transpiration. Transpiration is the loss of water vapour from the leaves of the plant, via the stomata. Water molecules stick to each other, and because of that the loss of water vapour pulls other water molecules behind it, and so moves water up the xylem from the root. It is rather like drinking from a straw, when sucking the end of the straw pulls the drink up in a continuous stream.

Different environmental conditions affect the rate of transpiration. Water is lost from the leaves by evaporation, which is, in effect, the diffusion of water molecules into the air. Therefore, anything that affects the rate of diffusion affects the rate of transpiration. Environmental factors that do this are as follows:

▶ **Temperature** – Increasing temperature speeds up the movement of particles, and so increases the transpiration rate as the water molecules move out more rapidly.

▶ **Humidity** – The rate of transpiration depends on the difference in concentration of water inside and outside the leaf. An increase in humidity decreases this concentration gradient, and so decreases the rate of transpiration.

Figure 5.14 Xylem cells showing adaptations for water transport.

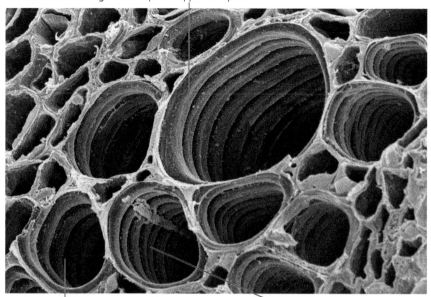

The cell walls are thickened with a substance called lignin, which makes them rigid and helps to support the plant.

The cells are dead and have no cytoplasm, which would get in the way when transporting water.

The cells have partially or entirely lost their end walls, so they form continuous tubes.

 Specified practical

Investigation into factors affecting transpiration

This experiment uses a piece of apparatus known as a potometer. There are many different designs of potometer, one of which is shown in Figure 5.15. The shoot should be cut and the apparatus assembled under water, to prevent air getting into the shoot or the potometer, because that causes an 'air lock' that stops the uptake of water.

When assembled, an air bubble is drawn into the capillary tube and, as water is absorbed by the shoot, the air bubble moves. The rate of movement gives an indication of how much water is being lost from the leaves of the plant.

Procedure

1 Design and carry out an experiment to investigate the effect of air movement on water loss in a plant. A hair dryer or fan could be used to create air movement.
2 Record your results and draw conclusions.

Questions

1 The potometer does not actually measure water loss from the leaves (that is, transpiration). It measures water uptake, and it is assumed that this is related to water loss (that is, the more water that is lost, the more water will be taken up). How accurate do you think this assumption is, and does it matter for the purposes of the experiment?

2 Now that you have carried out the experiment, are there any improvements that you could make to your method?

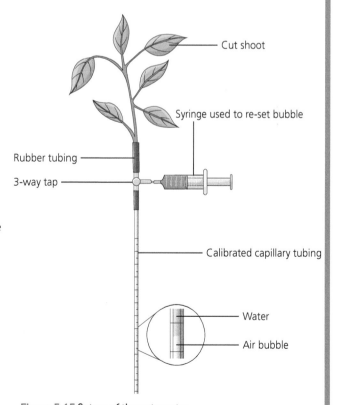

Cut shoot

Syringe used to re-set bubble

Rubber tubing

3-way tap

Calibrated capillary tubing

Water

Air bubble

Figure 5.15 Set-up of the potometer.

▶ **Wind or air movement** – Movement of air around the plant constantly blows away the water molecules on the outside of the leaves, preventing any moisture building up and maintaining a steep concentration gradient. So, increasing air movement or wind speeds up the rate of transpiration.

Light affects whether transpiration occurs or not, but does not affect its rate. In the dark, the stomata close and the rate of transpiration drops to almost zero. Once the stomata are open, increasing the light intensity has no effect.

▶ What does the phloem do?

Everywhere you find xylem cells in a plant, there are another set of tubes nearby, called phloem (Figure 5.16). The job of phloem tissue is to transport sugar (in solution) away from the leaves to other parts of the plant. Unlike xylem, which only transports water from the roots towards the leaves, phloem can move sugar solution in both directions, to all parts of the plant.

Two things can happen to this sugar.

▶ It can be used straight away to provide energy, by respiration. This is particularly important at the growing points of the plant, which are at the tips of the roots and shoots.
▶ If it is not needed immediately, the sugar is stored as starch, both in the leaves and in other parts of the plant.

Figure 5.16 Position of xylem and phloem in a stem and a root.

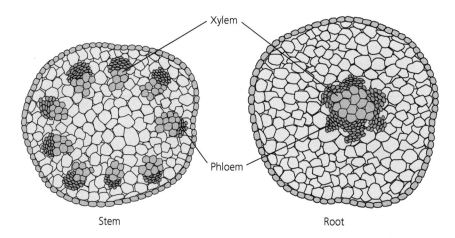

Xylem

Phloem

Stem

Root

▶ Why do plants need minerals?

Plants need a variety of minerals for healthy growth, but three of the most important are **nitrogen, potassium** and **phosphorus**. If a plant lacks one of these, it shows poor growth and specific symptoms.

▶ Lack of nitrogen (in the form of nitrates) causes generally poor growth, because this mineral is needed to make proteins for new cells.
▶ Lack of potassium is characterised by yellowing of the leaves.
▶ Lack of phosphorus (in the form of phosphates) results in poor root growth.

General-purpose fertilisers are often referred to as KPN fertilisers, because they contain all three of these elements – potassium (chemical symbol K), phosphorus (chemical symbol P) and nitrogen (chemical symbol N).

✔ Test yourself

11 Define the term 'transpiration'.
12 Plants lose most water on warm, dry, windy days. Explain why.
13 State two differences in the way water and sugar are transported around the plant.
14 When we want to see if a leaf has been carrying out photosynthesis, we test it for starch and not for sugar. Suggest a reason for this.
15 Why does a lack of nitrate result in poor growth of the plant?

Chapter summary

- Photosynthesis is the process whereby green plants and other photosynthetic organisms use chlorophyll to absorb light energy and convert carbon dioxide and water into glucose, producing oxygen as a by-product.
- The chemical reactions of photosynthesis within the cell are controlled by enzymes.
- Photosynthesis requires carbon dioxide, water and light, together with chlorophyll in chloroplasts to absorb the light.
- The rate of photosynthesis is affected by temperature, levels of carbon dioxide and light intensity.
- A limiting factor is one that is limiting the rate of photosynthesis at a given time.
- Temperature, levels of carbon dioxide and light intensity can act as limiting factors for photosynthesis.
- The occurrence of photosynthesis can be detected by testing a leaf using iodine in potassium iodide solution, which turns blue-black in the presence of starch.
- Glucose produced in photosynthesis can be respired to release energy, converted to starch for storage or used to make cellulose, proteins and oils.
- The leaf consists of the following structures: cuticle, epidermis, guard cells (which form stomata), palisade layer, spongy layer, xylem and phloem (which make up the veins).

- The lower epidermis of a leaf has pores called stomata (singular: stoma). The stoma is the pore itself, which is surrounded by, and opened and closed by, a pair of guard cells.
- Water is needed in plants for use in photosynthesis, for transport of sucrose and minerals and for support.
- Root hairs increase the surface area for absorption in a root.
- The uptake and movement of water through the root is a result of osmosis.
- Mineral salts are taken up by root hairs by active transport.
- The transport of water within plants occurs in the xylem.
- Transpiration is the loss of water vapour from the leaves and results in the movement of water through a plant.
- The rate of transpiration is affected by temperature, air movement and humidity.
- The phloem carries sugar from the photosynthetic areas to other parts of the plant for use in respiration or to be converted into starch for storage.
- Lack of nitrates in a plant results in poor growth; deficiency of potassium results in yellowing of the leaf; deficiency of phosphate results in poor root growth.
- KPN fertilisers are often used to ensure that plants have essential minerals.

▶ Chapter review questions

1 Describe the method involved in testing a leaf for the presence of starch.

Each of the stages involved in the method should be described in sequence and the reason for carrying out each stage should be included.

Your description must include reference to the colour changes shown by the leaf and what these changes indicate. [6]

(from WJEC Paper B2(H), Summer 2014, question 4)

2 a) Write the word equation for photosynthesis. (Do not use chemical formulae.) [1]

b) The graph below shows the rate of photosynthesis under differing environmental conditions of light and carbon dioxide.

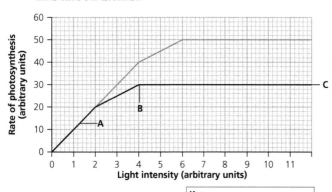

i) State why the rate of photosynthesis is low at point **A**. [1]

ii) Explain why the rate of photosynthesis has levelled off between points **B** and **C**. [1]

iii) State one way in which the rate of photosynthesis could be measured in the laboratory. [1]

iv) State one way in which the glucose produced during photosynthesis is used by the plant. [1]

(from WJEC Paper B2(H), Summer 2012, question 5)

3 The photograph below shows a tomato plant.

Some of the sugar made in photosynthesis is transported to the tomato fruits.

a) State the name of the tissue in plants that transports sugar. [1]

Siân grows tomato plants. She decides to use a fertiliser called Topgrow. The label from a bottle of Topgrow is shown below.

TOPGROW FERTILISER

CONCENTRATED NUTRIENT SOLUTION

Dilution: 1 part Topgrow: 200 parts water

Contents of bottle: 500 cm³

---

b) Using the instructions for use shown on the label, calculate the volume of diluted Topgrow that can be made from the contents of one bottle. [2]

c) Siân carried out a trial to find out the effect of using Topgrow on the tomato plants. She used tap water only on half the plants and diluted Topgrow on the rest. What else should Siân have done to make sure that the trial was a fair test? Give two suggestions. [2]

d) The table shows some of the results of the trial.

Treatment	Mean yield (mean mass of tomatoes per plant), in kg	Mean number of tomatoes per plant	Mean mass per tomato, g
Tap water	4.8	40	120
Topgrow	5.2	65	

i) Calculate the mean mass per tomato (in g) for Topgrow. [1]

ii) Siân was pleased with the effect of Topgrow on yield. Suggest why Siân was still disappointed with the results. [1]

e) Apart from nitrates, give the names of two other nutrients required for healthy plant growth. [2]

(from WJEC Paper B3(H), Summer 2013, question 1)

6 Ecosystems, nutrient cycles and human impact on the environment

 Specification coverage

This chapter covers the GCSE Biology specification section **1.6 Ecosystems, nutrient cycles and human impact on the environment** and GCSE Science (Double Award) specification section **1.6 Ecosystems and human impact on the environment**.

It covers the levels of organisation within an ecosystem, the principles of material cycling and issues surrounding sustainability. Opportunities are given to look in detail at the factors affecting communities and how the numbers of organisms and biomass within each level can be represented. The carbon cycle and nitrogen cycle are covered, along with how human activity affects them.

▶ Where do we get our energy from?

Key terms

Producer Living organism that makes food using light or chemical energy. Plants and some bacteria use light, while other bacteria use chemical energy.

Consumer Living organism that gets its energy by consuming food.

Herbivore Living organism that feeds entirely on plants.

Carnivore Living organism that feeds entirely on animals.

Omnivore Living organism that feeds on both plants and animals.

Decomposer Bacteria and fungi that break down the dead bodies of plants and animals.

Trophic level A group of organisms that occupy the same position in a food chain.

Energy arrives on our planet constantly in the form of sunlight. This energy passes from organism to organism by means of food chains. Plants are the first links in all food chains because they are producers – they change energy in sunlight into stored chemical energy. At this point, quite a lot of the energy is wasted – plants only manage to capture about 5% of the energy in sunlight. When plants are eaten by herbivores, some of the energy is passed to the next link, the consumers in the food chain. When the herbivore is eaten by a carnivore, the process of energy transfer is repeated. Energy passes in this way from carnivores to scavengers and decomposers, which feed on dead organisms. However, not all the energy stored by a herbivore is available to a carnivore that feeds on it. Much is used in life processes such as movement, growth, cell repair and reproduction. Some is lost in waste materials, and energy is also wasted as heat during respiration. Only the leftover energy is stored by the carnivore.

Consider the food chain through which energy flows when we eat fish such as tuna. Energy from sunlight is first used by plant plankton (microscopic algae). It then passes to animal plankton, then to small fish, then to larger fish, then to tuna and then to us. There is usually no predator to eat us, so we are the top carnivores in this food chain.

plant plankton → animal plankton → small fish → large fish

→ tuna → human

Specific names are given to each stage, or trophic level, in a food chain. Table 6.1 lists these, using the food chain above as an example.

Table 6.1 Stages in a food chain.

Organism	Name of stage
Plant plankton	Producer
Animal plankton	First stage consumer
Small fish	Second stage consumer
Large fish	Third stage consumer
Tuna	Fourth stage consumer
Human	Fifth stage consumer

This is actually an unusually long food chain. Most food chains do not reach beyond the third or fourth stage consumer, because the energy lost at each stage means there is often not enough left to support a fifth stage consumer. Note that an animal can operate at more than one stage. In the food chain shown, humans are a fifth stage consumer. When we eat fruit, for example, we are a first stage consumer and when we eat chicken we are a second stage consumer.

In nature, food chains often interlink, because most organisms eat a lot of different things and are eaten by many different animals, too. Interlinked food chains are called food webs. Figure 6.1 shows an example, but even this food web is an oversimplification of all the feeding relationships that would exist in this environment.

Discussion point

Often, the organisms get bigger as you go along the food chain. What are the reasons for this?

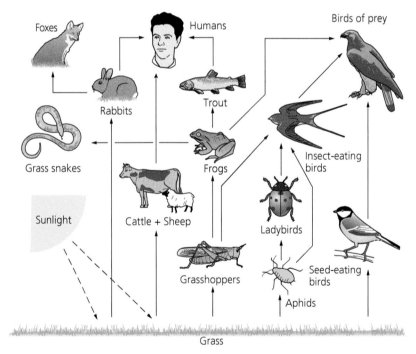

Figure 6.1 An example of a food web.

What do pyramids of numbers and biomass show us?

Feeding relationships can be illustrated as pyramids (Figure 6.2 and Figure 6.3). The width of each block in the pyramid is an indication of the number (or mass) of that type of organism at that feeding level.

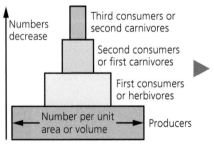

Figure 6.2 A pyramid of numbers for a grassland food chain.

Numbers decrease
Third consumers or second carnivores
Second consumers or first carnivores
First consumers or herbivores
Number per unit area or volume — Producers

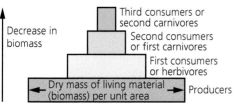

Figure 6.3 A pyramid of biomass for the same grassland food chain.

What do pyramids of numbers and biomass show us?

Key term

Dry mass The mass of a material after all the water has been removed.

Discussion point

Why is it better to use the dry mass of organisms when constructing a pyramid of biomass, rather than the wet mass? What is the disadvantage of using wet mass?

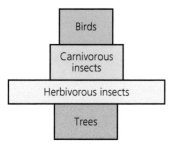

Figure 6.4 Example of a pyramid of numbers that is the 'wrong' shape.

These pyramids can tell us more about the energy that is available to organisms living in a measured area or volume. The pyramids can be drawn in different ways:

▶ a **pyramid of numbers** shows the number of organisms per unit area or volume at each feeding level
▶ a **pyramid of biomass** shows the **dry mass** of organic material per unit area or volume at each feeding level.

Pyramids of biomass give a more accurate picture than pyramids of numbers. Pyramids of numbers are sometimes not actually pyramid shaped. Look at Figure 6.4. It is pretty clear what has happened here. Unlike in many food chains, here the producers (trees) are much larger than the insects that feed on them. One tree can support thousands of insects, so in a pyramid of numbers the bottom block, representing producers, is narrower than the block for the first stage consumers. However, a tree weighs a lot more than all the insects feeding on it put together, so a pyramid of biomass will be pyramid shaped as expected.

Activity

Calculating the efficiency of energy transfers in a food chain

The amounts of energy taken in by organisms at different stages of a food chain were calculated as shown in Table 6.2 below.

Table 6.2 Energy taken in at each stage of a food chain.

Stage	Total energy, in kJ
Producers	97 000
First stage consumers	7 000
Second stage consumers	600
Third stage consumers	50

The efficiency of energy transfer at any stage can be calculated as follows:

$$\text{efficiency} = \frac{\text{energy in later stage}}{\text{energy in earlier stage}} \times 100\%$$

Questions

1 Calculate the efficiency of energy transfer for each stage in the food chain.
2 At each stage, the efficiency is quite low. Suggest reasons for this.
3 Using the data, estimate the energy that would be contained in a fourth stage consumer.
4 Suggest why it is unlikely that this food chain could have a fifth stage consumer.

Test yourself

1 What is the source of energy in every food chain?
2 What is another name for a first stage consumer?
3 Why is it impossible to say what type of consumer a human is?
4 Why is a pyramid of biomass more accurate than a pyramid of numbers?

▶ Why do we need microorganisms?

We often think of microorganisms as harmful or a nuisance. They can make our food go off or cause disease. However, microorganisms play a vital role in life on Earth. They get rid of animal wastes and the dead bodies of animals and plants and, in doing so, return vital nutrients to the soil, which can then be used for new growth. Examples include nitrates, which are formed from the breakdown of proteins in dead tissues and used by plants to form new cells. Phosphates, which are needed for many vital functions in living bodies, are also recycled. The cycling of these nutrients ensures a balance in ecosystems, as the processes that remove materials are balanced by the processes that return them.

> **Key term**
>
> Ecosystem The living organisms in an area together with the non-living components of their environment (for example, soil, climate), interacting as a system.

▶ How is carbon recycled?

It could be claimed that carbon is the most important element of all for life, as all life on our planet is carbon based. There is a fixed supply of carbon on Earth, and life has to make do with this amount, as no extra comes in from outside. Therefore, it is essential that the carbon on the planet is recycled and constantly renewed. The way this happens is shown in Figure 6.5.

Figure 6.5 The carbon cycle.

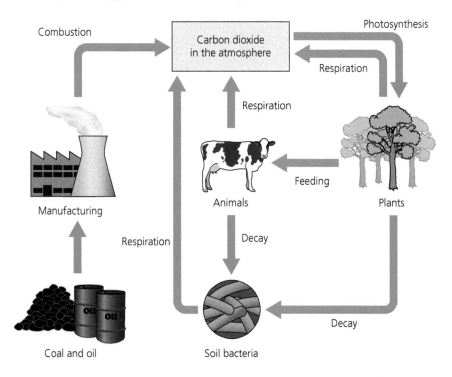

Carbon dioxide in the air is made into food by green plants in photosynthesis. Animals get their carbon by eating plants (or other animals). The carbon in dead animals and plants is released back into the atmosphere by the process of decay. The bacteria involved release carbon dioxide when they respire. Living animals and plants also respire, and so put carbon dioxide back into the atmosphere.

Fossil fuels were made from the dead bodies of plants and animals millions of years ago. They were not completely decomposed, so a lot of the carbon in them was 'locked' into the fossil fuels. When fossil fuels are burnt, the carbon in them is released as carbon dioxide, adding to the levels in the atmosphere. Humans have only started extracting and burning fossil fuels in large quantities in the last 200 years. This has disrupted the balance and led to an increase in the levels of carbon dioxide in the atmosphere.

Microorganisms play a role in the carbon cycle because, like all living things, they respire. Their role in the nitrogen cycle is much greater, however.

How is nitrogen recycled?

The nitrogen cycle is shown in Figure 6.6. Nitrogen in the air is changed by nitrogen-fixing bacteria in the soil into **nitrates**, which plants can absorb and use. Nitrogen-fixing bacteria are also found in the roots of one group of plants, the legumes (peas, beans and clover), in special structures called root nodules. The nitrates absorbed by the plants are passed on to animals that eat the plants, and the nitrogen is eventually returned to the soil in urine and faeces from animals, and when dead animals and plants decay. The nitrogen from wastes and decay is in the form of **ammonia**, which cannot be used directly by plants. Bacteria in the soil convert the ammonia into nitrates, however, which are then absorbed by plants. Nitrogen is returned to the air by denitrifying bacteria.

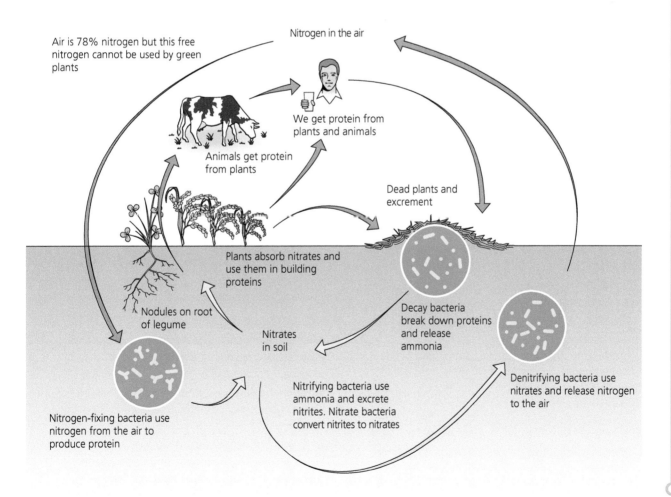

Figure 6.6 The nitrogen cycle.

Sometimes, in agricultural land, the cycle gets out of balance and the level of nitrogen in the soil drops. This can happen for two reasons:

▶ The density of plants can be very high, so that a lot of nitrogen is removed from the soil.
▶ The plants that grow in the fields are harvested and removed. As they do not die in the fields, the nitrogen in them never gets returned to the soil.

On farms this imbalance has to be restored by the addition of nitrogen-containing fertilisers, either natural or artificial, to the soil.

🧪 Practical

How does urine help keep the nitrogen cycle going?

As part of the process of recycling nutrients, decomposers secrete enzymes into the soil that break down waste such as urea. One of these enzymes is called urease. This experiment investigates the effect of the concentration of urease on its reaction with urea. Urease catalyses (speeds up) the breakdown of urea to release ammonia. The ammonia dissolves in water to form an alkaline solution, which can be converted in the soil into nitrates.

Safety notes

Eye protection should be worn, and spillages onto skin should be washed off immediately.
Take care to avoid scalding with the high temperature water bath.

Apparatus

> 4 test tubes
> test-tube rack
> labels
> 3 syringes
> dropping pipette
> Bunsen burner, tripod and gauze
> large beaker (water bath)
> pH colour chart
> 0.1 mol/dm³ urea solution
> fresh urease solution
> boiled and cooled urease solution
> 0.1 M hydrochloric acid
> Universal indicator solution
> distilled water
> stopclock or stopwatch
> eye protection

Procedure

1 Wear eye protection. You will need a boiling water bath for this experiment. Half fill the large beaker with water and start to heat it now.
2 Label four test tubes A, B, C and D.

3 Using a syringe, add 3 cm³ of urease solution to tube D and place the tube in the boiling water for 4 minutes. Remove the tube and allow it to cool. (You can speed up the cooling process by running the bottom of the tube under cold water.)
4 Using the second and third syringes, add the following to each of the tubes:
> 5 cm³ of urea solution
> drops of hydrochloric acid
5 Using a dropping pipette, add 10 drops of Universal indicator solution to each tube.
6 Using the first syringe, add urease solution to tubes A, B and C as follows:
> Tube A – 1 cm³ urease
> Tube B – 3 cm³ urease
> Tube C – 5 cm³ urease
Tube D already contains 3 cm³ of boiled and cooled urease.
7 Draw up a suitable table to record your results.
8 Shake the tubes, compare the colour with the pH chart and record the approximate pH of the contents of each of the tubes in the table.
9 At two-minute intervals, shake the tubes and record the approximate pH of each tube (A, B, C and D). Continue the readings for 12 minutes.
10 Collect the results of other groups in the class to act as repeat readings.
11 Display your results as a line graph.

Analysing your results

1 What are your conclusions from this experiment?
2 How certain are you of these conclusions? Give reasons for your answer. (Hint: look at variations between the different groups' results.)
3 Why do you think hydrochloric acid was added to the tubes?
4 What was the point of tube D?
5 In this experiment, the urease used was the same concentration throughout, but different volumes were used in each tube. Does this mean the experiment is not valid as a means of testing the effect of concentration?

5 Why do chemical elements necessary for life need to be recycled?
6 What living process releases carbon into the atmosphere?
7 What role do nitrogen-fixing bacteria play in the nitrogen cycle?
8 Why do nitrogen levels in farmland drop over time, whereas those in wild habitats tend to remain stable?
9 How is nitrogen lost from living animals?

▶ How do we balance the needs of humans and wildlife?

Most people think that conserving wildlife is a good thing to do. The issue can be complicated when conservation measures mean a significant disadvantage for the human population, or when conserving one species can harm others. People need housing, but building it can destroy animal and plant habitats. People need food, but converting wild habitats into farmland alters and reduces the numbers of species that live there. The world needs to build up alternative sources of energy, but the development of structures such as tidal barrages and hydroelectric power stations can destroy or completely alter natural habitats. It is also worth remembering that over the whole of history, environments have constantly changed – it is not 'natural' for everything to stay the same.

Compromise is always necessary when the interests of wildlife and humans conflict. No two situations are the same. For instance, a human need for a new leisure facility is not as important as creating farmland to help feed a starving population. Development of an area that would destroy the habitat of a rare species is different from a similar development of land where the species involved are common and relatively few.

▶ Do we want cheap food or happy farm animals?

Battery farming, where animals are kept in huge numbers in a small space, is an environmental issue that some people have strong opinions about (Figure 6.7). It is just one example of

Figure 6.7 Battery-farmed chickens are kept in very confined spaces.

intensive farming. Intensive farming is an agricultural system that aims to produce a maximum yield from the land available. It applies to both animals and plants, and – as well as battery farming – it involves the use of chemicals such as pesticides and fertilisers to increase yield and to control diseases.

There are advantages and disadvantages of intensive farming, and scientific evidence can clarify these, but in the end people have to decide their own opinions about it. Scientific data cannot decide an ethical issue, but can provide the accurate information people need to help them make their decisions.

The advantages of intensive farming are:

▸ The yield is high because more livestock animals or crop plants can be kept and conditions can be controlled. Therefore, the food is cheaper to produce and more profitable for the farmer, which may keep him or her in business. If farms go out of business, the UK becomes less self-sufficient in food.
▸ Food is cheaper in the shops, allowing people to choose a healthier diet, even on a tight budget.
▸ By increasing the yield, it allows the UK to grow more food, to meet the needs of a growing population.

The disadvantages are:

▸ The chemicals used (for example, pesticides and antibiotics used to control disease in livestock) could enter the human food chain and get into our bodies.
▸ The chemicals can cause pollution and harm wildlife other than pests.
▸ Natural environments are destroyed. For example, hedgerows are uprooted to make large fields suitable for intensive farming.
▸ Although no one can really know what an animal 'feels', it is likely that intensive farming causes animals stress and discomfort. Their quality of life is very poor.

▸ What pollutants are in our environment?

A pollutant is something that has been added to the environment and which damages it in some way. There are many types of pollutant. Pollutants are not necessarily 'unnatural', as some natural substances can be harmful if introduced in the wrong place or in large quantities. Some common pollutants are:

▸ solid or liquid chemicals, such as oil, detergents, fertilisers, pesticides and heavy metals
▸ gaseous chemicals such as carbon dioxide, methane, chlorofluorocarbons (CFCs), sulfur dioxide and nitrous oxides
▸ human and animal sewage
▸ noise
▸ heat
▸ non-recyclable household waste.

It is impossible to prevent pollution, but we must try to limit the levels of pollution so that they do little or no permanent damage to the environment.

Activity

How can we tell how polluted the environment is?

Some pollutants can be measured directly, and water pollution can also be detected by a fall in the **oxygen level** or a change in **pH** in streams and rivers. Scientists can often judge the overall level of pollution by using indicator species (Table 6.3). Some plants and animals are more tolerant of pollution than others. In any environment, there are certain species you would expect to find. If some expected species are absent, this can give an idea of how polluted the environment is.

Table 6.3 Species indicators of pollution in streams.

Water quality	Species present
Clean water	Stonefly nymph, mayfly nymph
Low level of pollution	Freshwater shrimp, caddis fly larva
Moderate pollution	Water louse, bloodworm
High pollution	Sludge worm, rat-tailed maggot

Figure 6.8 shows a stream that scientists were studying for pollution. Two farms, Mill Farm and Tipton Farm, were near the stream and it was thought that sewage from one or both farms might be getting into the stream. The stream was sampled at five places, labelled A–E on the diagram.

The results of the study are shown in Table 6.4.

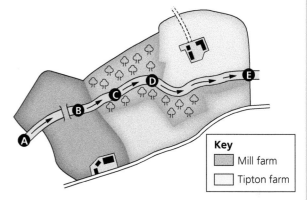

Key
Mill farm
Tipton farm

Figure 6.8 Map of the study area.

Table 6.4 Results of a study assessing pollution levels in a stream, using indicator species.

Sample point	Numbers of each species found, per m^2							
	Stonefly nymph	Mayfly nymph	Freshwater shrimp	Caddis fly larva	Bloodworm	Water louse	Rat-tailed maggot	Sludge worm
A	11	15	5	12	0	2	0	0
B	0	0	3	4	6	16	12	3
C	0	0	3	8	8	14	2	0
D	0	0	4	10	4	6	0	0
E	0	0	0	4	12	20	2	0

From the results, write a report of the pollution of the stream and its likely causes. Make sure your report is detailed and use evidence from the results to justify your conclusion.

▶ Why do we worry about chemicals 'entering the food chain'?

The main chemicals to be concerned about here are **heavy metals** and **pesticides**. Living things need small quantities of metals but too much can harm them. Some metals, such as lead and mercury, are poisonous even in small quantities. Most heavy metal pollution is caused by industrial processes. There used to be a lot of lead pollution from vehicles burning petrol, but nowadays petrol is lead-free, although leaded petrol is still used in some developing countries.

Pesticides are poisonous chemicals that are used to kill agricultural pests, usually by spraying on crops. Some of them take

How polluted is the air you are breathing?

Lichens are used as indicators of air pollution. Figure 6.9 shows lichens that are found in clean air and lichens that are found in areas polluted with nitrous oxides or sulfur dioxide. It also shows examples of plants that are not lichens, but are sometimes mistaken for them.

All these lichens grow on the bark of trees. Carry out a survey of the trees around your school to see if the air in your area is polluted.

> If you were comparing your area with another one, explain why it would be necessary to look at just one species of tree in both areas in order to make the test fair.

> If you wanted to compare your area with others, it would be useful if you could get quantitative data (numbers) to compare. How could you design your survey so that it gave some sort of figure for the air pollution that could be reliably compared with another area?

Nitrogen-loving

Xanthoria parietina

Physcia tenella

Usnea cornuta

Sulfur-tolerant

Xanthovia polycarpa

Hypogymnia physodes

These are not lichens!

Moss

Desmococcus

Figure 6.9 Lichens and air pollution indicator species.

time to break down, so traces of them can be found on fruits and vegetables in the shops. Pesticides left in the soil can also be washed into rivers and streams by rainfall, and pesticide sprays can drift in the air beyond the area being sprayed.

In the UK there are controls on the use of these polluting chemicals. Some are banned in certain situations and the Environment Agency monitors the environment to look for any signs of harmful levels of pollutants. Although there are occasional accidents, causing high levels of pollution, the levels of these pollutants released into the environment are usually small and controlled. Problems arise, though, if these chemicals 'enter the food chain'.

A famous case of this occurred in Minemata, Japan in the 1950s. The city is on the shore of Minemata Bay, and the population lived almost entirely on fish caught in the bay. Many people in Minemata suddenly started showing the symptoms of mercury poisoning, and 20 died. Mercury was used in a factory on the edge of the bay, but there had been no large spillage. However, the mercury had been absorbed by microscopic plants in the bay, which were part of the human food chain. Part of the food web in the bay is shown in Figure 6.10.

Figure 6.10 Part of the food web around Minemata Bay. The humans ate a variety of other fish and shellfish apart from mullet. These in turn all ate the microscopic plants and animals.

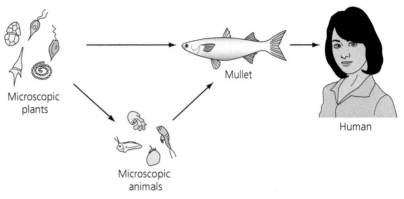

The poisoning had happened in this way:

1 The microscopic plants absorbed mercury in the water.
2 The microscopic animals ate large quantities of the plants, and the mercury built up inside them.
3 The fish ate very large quantities of the microscopic plants and animals, and so the mercury built up to even higher levels in them.
4 The fish were, in effect, poisonous because of the levels of mercury they contained. When the humans ate a lot of these fish, the mercury levels in them became so high that it made them very ill or killed them.

Organisms at the top of any food chain will accumulate the highest levels of any poison that enters the chain. As humans are at the top of all the food chains we occupy, we are particularly at risk.

▶ Why do sewage and fertilisers kill fish?

Sewage and fertilisers sometimes get into streams and rivers from farmland, washed from the soil by rain. This starts a process called eutrophication, which can kill fish and other animals. It happens in the following way:

1 The sewage or fertiliser causes an increase in the growth of microscopic plants.
2 These plants have short lives so, soon afterwards, the number of dead plants in the water goes up.
3 Bacteria rot the bodies of the plants, and because there are so many dead plants the population of bacteria goes up sharply.
4 These bacteria use oxygen for respiration, so the oxygen level of the water goes down.
5 Animals such as fish, which need a lot of oxygen, die because there is now not enough oxygen in the water.

Key term

Eutrophication The process by which pollution by fertilisers results in lowered oxygen levels in bodies of water.

In small bodies of water such as ponds there is another problem. The microscopic plants may grow so much that they form a complete blanket over the surface, so cutting off the light that the plants at the bottom of the pond need to survive.

✔ **Test yourself**

10 Define the term 'pollutant'.
11 Why, in general, are organisms at the top of a food chain most at risk from pollutants?
12 What is an indicator species?
13 Which organisms are responsible for the lowered level of dissolved oxygen seen in eutrophication?

⬇ **Chapter summary**

- Food chains and food webs show the transfer of energy between organisms and involve producers, first stage consumers (herbivores), second and third stage consumers (carnivores), and decomposers. Some food chains involve fourth or even fifth stage consumers, but this is unusual.
- At each stage in the food chain energy is used in repair and in the maintenance and growth of cells, while energy is lost in waste materials and respiration.
- Pyramids of numbers and biomass indicate the numbers or mass of the organisms at different trophic levels.
- The efficiency of energy transfers between trophic levels affects the number of organisms at each trophic level.
- Microorganisms, bacteria and fungi are important in causing decay. They feed on waste materials from organisms, and when plants and animals die their bodies are broken down by microorganisms.
- Microorganisms respire and release carbon dioxide into the atmosphere.
- Nutrients are released in decay – for example, nitrates and phosphates – and these nutrients are then taken up by other organisms, resulting in nutrient cycles.
- In a stable community, the processes that remove materials are balanced by processes that return materials to the environment.
- Carbon is constantly cycled in nature via photosynthesis, which incorporates it, and respiration, which releases it; burning fossil fuels releases carbon dioxide.

- Nitrogen is also cycled, through the activity of soil bacteria and fungi that act as decomposers, converting proteins and urea into ammonia. Ammonia is then converted to nitrates, which are taken up by plant roots and used to make new proteins.
- Nitrogen fixation is the process by which nitrogen from the air is converted to nitrates.
- Levels of nitrates in the soil can be depleted by modern farming methods.
- Human requirements for food and economic development need to be balanced with the needs of wildlife.
- Intensive farming methods – such as using fertilisers, pesticides, disease control and battery methods to increase yields – have both advantages and disadvantages.
- Indicator species and changes in pH and oxygen levels may be used as signs of pollution in a stream and lichens can be used as indicators of air pollution.
- Some heavy metals, present in industrial waste and pesticides, enter the food chain, accumulate in animal bodies and may reach a toxic level.
- Untreated sewage and fertilisers may run into water and cause rapid growth of plants and algae, which then die and are decomposed. The microbes that break them down increase in number and their respiration can use up the oxygen in the water, causing harm to animals living there. This is known as eutrophication.

▶ Chapter review questions

1 Some organisms living in a large lake and their total biomass in kg are shown below. They are not drawn to scale.

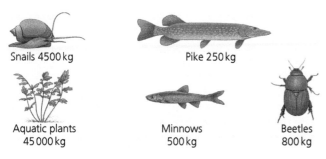

Snails 4500 kg

Pike 250 kg

Aquatic plants 45 000 kg

Minnows 500 kg

Beetles 800 kg

a) Which of the organisms above are likely to be present in the smallest numbers? [1]

The organisms above all form part of the same food chain.

b) Draw a labelled diagram to show a pyramid of biomass containing all of these organisms. [2]

The pike in the lake are affected by a parasite, called a fish louse, which lives on their skin. There would be many of these parasites on each pike but their biomass would be less than the biomass of the pike.

c) How would you add this information to the pyramid you drew in (b). Choose from the following: [1]

A Place them at the tier above the pike.

B Place them at the bottom of the pyramid.

C Place them below the minnows.

D Place them in the tier below the pike.

d) Explain how a pyramid of numbers, for some organisms living on land, could look like the one shown below: [2]

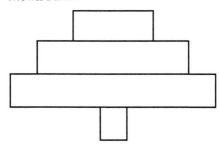

(from WJEC Paper B1(H), Summer 2014, question 1)

2 The Welsh Environmental Authority (Natural Resources Wales) monitors the concentration of nitrate and fish populations in rivers and lakes. The graph shows the results for a lake near farmland, which was regularly treated with fertiliser from 1980 to 1990.

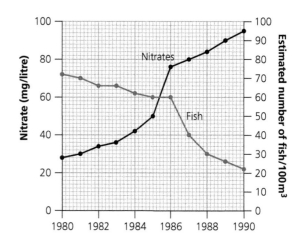

Use the graph and your knowledge of indicator species and pollution by fertilisers to answer the following.

a) Between which years was the highest rate of increase of nitrate pollution in the lake? [1]

b) In which year would you expect the highest concentration of: [1]

i) oxygen in the lake

ii) bacteria in the lake?

c) Sketch a line graph to show the expected changes in the biomass of plants in the lake from 1980 to 1990. [3]

(from WJEC Paper B1(H), Summer 2014, question 6)

7 Classification and biodiversity

🏠 **Specification coverage**

This chapter covers the GCSE Biology specification section 2.1 Classification and biodiversity and GCSE Science (Double Award) specification section 4.1 Classification and biodiversity.

It covers an overview of the need for classification and how different organisms show adaptations that enable them to compete successfully for resources within their habitat. Biodiversity is also covered, along with factors that affect biodiversity and how it can be measured.

▶ How do living things vary?

To answer this question fully would take a very long time. Living things show a huge range of variations, even within a single species. Features vary, as does size and level of complexity.

The smallest living organism is probably the bacterium *Mycoplasma gallicepticum*, which measures 200–300 nanometres (a nanometre is one millionth of a millimetre). Viruses are even smaller than that, but many scientists do not classify viruses as living things. The largest living thing so far discovered is a type of mushroom, *Armillaria solidipes* (Figure 7.1). A colony of this fungus was discovered in the Blue Mountains in Oregon, USA in 1998, and was found to measure 3.8 kilometres across. The fungus is mostly underground with just the mushrooms coming above the surface, so its size had gone unnoticed.

Figure 7.1 *Armillaria solidipes* – this mushroom species includes the largest living organism on the planet.

Many people might think of the blue whale as the largest living thing, but it is only the largest animal, growing up to 24 metres long and weighing up to 190 tonnes (Figure 7.2). You might wonder how you weigh a whale. They are too big to fit on any known weighing scale, but dead whales have been weighed in bits and then the figures added up!

Figure 7.2 The blue whale is the largest animal on Earth.

Living organisms also vary greatly in complexity, from just a single cell to organisms that have trillions of cells, arranged into tissues, organs and organ systems. It is estimated that an average human has about 70 trillion cells.

► How can we organise the huge variety of living things?

It is very difficult to know how many species there are on Earth (new ones are constantly discovered and others become extinct) but the latest estimate in 2011 put the figure at 8.7 million (give or take 1.3 million!). In order to study this huge number of species, they must be put into more manageable groups. In general terms, plants could be divided into non-flowering and flowering varieties, and animals into vertebrates (with a backbone) and invertebrates (without a backbone). This is not how scientists group living things, however. They have a more complex and detailed system with many more groups, each group having similar features. For example, mammals (the group humans belong to) all have hair and feed their young on milk.

Earlier in the chapter, the largest organism in the world was named as *Armillaria solidipes*. This is its scientific name. All species have a scientific name, even though some have a 'common' name as well. The scientific name, which always consists of two words, is used by scientists throughout the world. This means that when the name is used, everyone in the scientific community knows which organism is being referred to. Common names vary in different languages (and even in different regions of the same country) and so using them could cause confusion. The woodlouse (Figure 7.3 on the next page), for example, has many different names in different parts of the UK, including monkey pea (Kent), cheeselog (Berkshire), slater (Scotland), granny grey (South Wales) and parson's pig (Isle of Man)! What's more, all of these names are used for all types of woodlouse, of which there are 35 different species in the UK.

Figure 7.3 This is *Oniscus asellus* – also known as woodlouse, granny grey and monkey pea, depending on where you come from!

▶ How are organisms adapted to their environment?

One reason why even closely related species can show quite a lot of differences is that, through evolution, species become adapted to their environment. Features develop that help the organisms to survive, and if two closely related species live in different environments, they will adapt in different ways. These adaptations are of two sorts:

▶ **Morphological adaptations** are structural adaptations of the organism (either internal or external) – for example, colour of fur, leg length, petal shape, reduced size of appendix and so on.
▶ **Behavioural adaptations** could include the time of day when an animal is active or the type of food it eats. Plants have very limited 'behaviour', so this mostly applies to animals.

Some examples of adaptation to the environment are shown in Figures 7.4 and 7.5.

Leaves are reduced to spines to reduce water loss

Stem is fleshy and stores water

Long root system to access water deep underground

Figure 7.4 The cactus has quite extreme adaptations because its desert environment is also extreme.

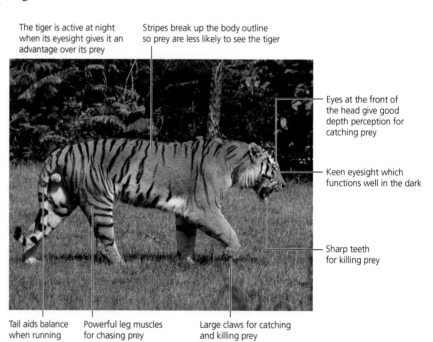

The tiger is active at night when its eyesight gives it an advantage over its prey

Stripes break up the body outline so prey are less likely to see the tiger

Eyes at the front of the head give good depth perception for catching prey

Keen eyesight which functions well in the dark

Sharp teeth for killing prey

Large claws for catching and killing prey

Powerful leg muscles for chasing prey

Tail aids balance when running

Figure 7.5 The tiger's adaptations not only suit it to its way of life as a predator, but also to its environment (it is successful in the jungle, but its stripes would provide much less effective camouflage in grassland).

→ Activity

Adaptation to the environment

Table 7.1 Examples of animals living in different environments.

Environment	Conditions	Examples
Desert	Extreme heat Shortage of water	Kangaroo rat Camel Desert iguana
Arctic	Extreme cold Covering of snow and ice	Polar bear Caribou Arctic hare
Savannah	Hot Open, so easy for predators and prey to be seen by each other Dry for most of the year except for wet season Food shortage in the dry season	Zebra Lion Meerkat

For each environment listed in Table 7.1, research one animal (either from the examples provided in the table or choose your own). For each animal, find and record one **morphological adaptation** and one **behavioural adaptation** to its environment.

▶ What resources do organisms need?

In order to survive, all organisms need a supply of energy. Plants get their energy directly from sunlight by photosynthesis; animals have to get their energy from food (that is, from other living organisms). Energy is needed to carry out all living processes but a supply of raw materials is also needed for chemical processes and to build bodies. Any environment that is to sustain life needs an adequate supply of these materials. Energy constantly enters most ecosystems in the form of sunlight, but the raw materials are limited and have to be recycled over and over again.

The resources needed by living things are summarised in Table 7.2.

Table 7.2 Resources needed by living things.

Resource	Needed by
Light	Plants to make food for energy
Food	Animals for energy
Water	All living organisms, for all the chemical reactions that take place in cells
Oxygen	All living organisms that respire aerobically, to break down food and release its energy
Carbon dioxide	Plants for photosynthesis
Minerals	All living organisms – specific minerals are needed for particular chemical reactions that take place in cells

▶ What happens when resources are limited?

There is always a limited amount of energy coming into an ecosystem and a limit to the other resources available. This puts a limit on the number of living things an ecosystem can support.

The organisms have to compete with each other for the resources and those that are better at this competition will survive better than the rest. Competition always takes place between members of the same species, because they all require the same things (for example, they eat the same food), but it also occurs between organisms of different species with similar needs. Competition puts a limit on the potential size of a population, although other factors are also important – for example, **predation**, **disease** and **pollution**.

Competition means that one or more resources are limited, and so the population cannot reach a size that it might have done if the resources were infinite (which they never are). Predation, disease and pollution all contribute to the death rate in a population, and so obviously limit its size.

→ Activity

Competition in flour beetles

Scientists kept two very similar species of flour beetles (the red flour beetle and the confused flour beetle) in some flour, which provides them with both food and a habitat. They found that after about 350 days, the red flour beetles had died out, leaving only the confused flour beetles. They also kept beetles of each species on their own in separate samples of flour. Some of their results are shown in Figure 7.6.

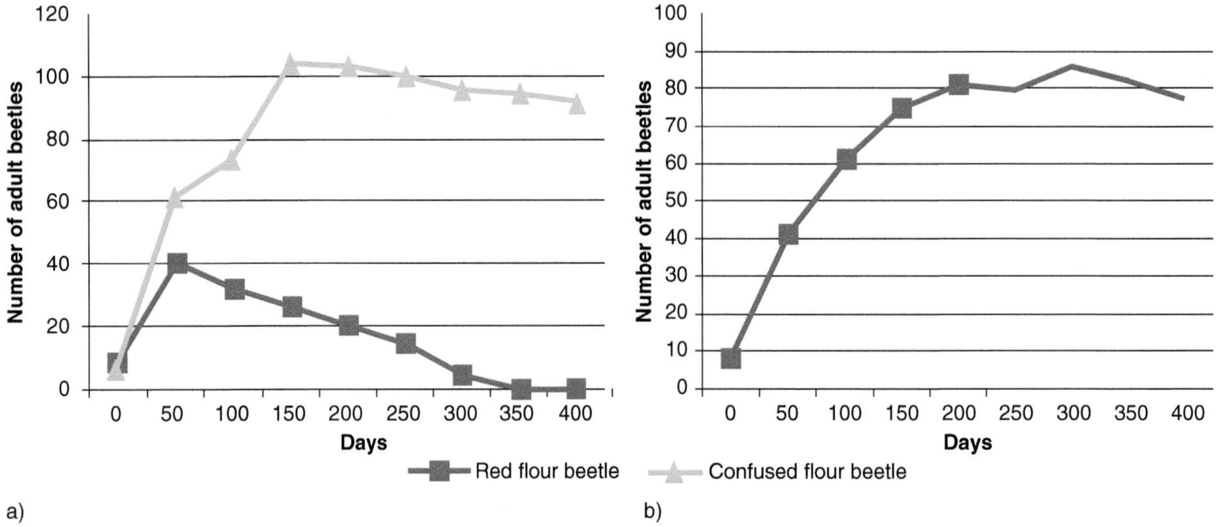

a)   b)

Figure 7.6 Competition in flour beetles. a) Results with both beetles together, b) results with red flour beetle alone.

Questions

The scientists concluded that the flour beetles were competing for a resource, and that the confused flour beetle was more successful in this competition than the red flour beetle.

1 Why was it important for the scientists to study how the red flour beetle population grew when it was on its own?
2 What resource(s) might the two beetle populations be competing for?
3 How strong is the evidence that there is competition between the two species of beetle? Explain your answer.
4 Suggest a reason why both populations increased in the first 50 days.
5 The scientists kept the temperature of the flour the same throughout the experiment. Suggest why they did that.

Practical

Competition in cress seeds

This experiment looks at the effect of competition for water on the germination of cress seeds.

Apparatus

> 2 sheets of filter paper
> 2 Petri dishes
> syringe
> cress seeds
> forceps

Procedure

1 Prepare two sheets of filter paper. Cut each so that it fits inside the lid of a Petri dish. Draw a grid on each piece of paper, as shown in Figure 7.7.

2 Put each piece of filter paper into a Petri dish.

3 Using a syringe, add just enough water to dampen all of the filter paper in one Petri dish. Note how much water you used.

4 Add the same amount of water to the second Petri dish.

5 Using forceps, carefully place one cress seed inside each square on the filter paper in the first Petri dish. Try to put the seed in the middle of the square. Count and record the total number of seeds used.

6 Put three seeds into each square in the second Petri dish. Count and record the total number of seeds used.

7 Put the lids on the Petri dishes and leave for 3–4 days for the seeds to germinate.

8 Record the percentage germination of the seeds in each dish.

9 Draw conclusions and explain your results.

10 Stephanie and Ivan did the experiment. Stephanie thought the experiment should be repeated, but Ivan said that because they had used lots of seeds, they had already repeated it. Discuss the merits of each idea.

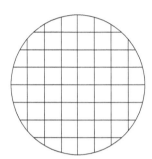

Figure 7.7 A grid of lines 1 cm apart, drawn on a filter paper circle (not to scale).

 Test yourself

1 Why do scientists give organisms scientific names?

2 What is a morphological adaptation?

3 Why do both animals and plants need light, even though only plants use it directly?

4 Why will there always be competition between members of the same species living in the same area?

5 Apart from competition, name two other factors that can limit the size of a population.

▶ What is biodiversity, and why is it important?

Biodiversity is the number of different species (of all types) in a particular area. It relates not only to the number of animals and plants, but also their variety. The 'area' concerned is of no fixed size – you could talk about the biodiversity on a sea shore, or in Wales, or in Europe, and so on.

Biodiversity is a good thing, because it leads to stable environments that can resist potentially harmful situations. Let's

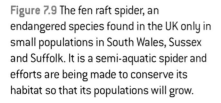

Figure 7.8 A food chain and a food web.

look at a type of environment that often has low biodiversity – a large field growing only one type of crop. Imagine that the crop is eaten by just one species of insect, and that the insect is eaten by just one species of bird. (No environment would ever be that simple, but this example is just to show the principles in a simple way.)

In this environment, there is just one food chain, as shown at the top of Figure 7.8.

Now imagine that the farmer uses an insecticide to kill most of the insects. The birds will have nothing to eat and will go elsewhere, where they can get food. The few insects that survived will now not be eaten by anything, and their population will grow again very fast, causing severe damage to the crop, before eventually the birds will return, in response to the return of their food. A change in the population of one species can therefore have big effects on the others.

Now, let's consider a more complex environment with more species living in it, forming a food web at the bottom of Figure 7.8.

Suppose that the farmer kills many of insect species D with insecticide. This time, insect E, which does not feed on the crop, can supply food for bird P. Bird Q might now have less food to eat, but can still survive by eating insect F. All of the species can remain in the area, even if their numbers alter a bit. The environment is more **stable**.

In reality, there are always many more species in an environment than in these examples, but the principle holds true – the greater the biodiversity, the greater the stability of the environment.

Around the world, efforts are being made to preserve biodiversity, and to save endangered species. A lot of publicity is given to large animals, but biodiversity depends on keeping as wide a variety of species as possible, and it is therefore important to conserve plants, worms, insects, spiders and so on, as well (Figure 7.9).

Figure 7.9 The fen raft spider, an endangered species found in the UK only in small populations in South Wales, Sussex and Suffolk. It is a semi-aquatic spider and efforts are being made to conserve its habitat so that its populations will grow.

Figure 7.10 The Soay sheep is an example of an ancient breed being conserved, whose genes may prove useful in the future.

Biodiversity is also helpful in other ways. Hundreds of years of selective breeding in domestic and farm animals and crops have sometimes resulted in the loss of resistance to certain diseases. It is important we do not let the ancient breeds become extinct in case we need to strengthen current breeds, or re-introduce disease resistance, by cross-breeding them with ancient species in the future (Figure 7.10).

▶ How can biodiversity be maintained?

The first problem to be tackled if biodiversity is to be maintained is to have a measure of it, and to repeat the measurement at intervals so that any changes can be identified. In the UK, a group called the UK Biodiversity Partnership collates data and assesses the biodiversity in the country by monitoring a set of 18 biodiversity indicators. Some data from the UK Biodiversity Partnership report for 2010 is shown in Figure 7.11 in the activity below.

Once accurate data for a population has been collected over a period of time, it is possible to use a mathematical model to predict what will happen to the population in the future, and to highlight possible future problems.

There are various ways in which biodiversity can be maintained, locally or nationally:

- ▶ breeding and release programmes to boost populations
- ▶ active conservation of habitats of threatened species
- ▶ creation of habitats that have declined (planting, landscaping, and so on)
- ▶ control of invasive species that may be spreading and pushing out other species
- ▶ legislation to protect habitats or individual species
- ▶ controlling pollution or other factors that might be threatening species or their habitats.

Legislating (that is, making laws) to protect habitats or species can be difficult in some cases, as wildlife needs may conflict with human needs. In addition, the living world is very varied and it can be quite difficult to frame legislation that will cover all situations fairly.

→ | Activity

Biodiversity in the UK

The graph shows data from the UK Biodiversity Partnership report for 2010.

Over the study period, populations of seabirds have increased, water and wetland bird populations have been more or less stable, but there has been a decline in numbers of woodland and farmland birds.

Questions

1 Suggest a possible reason for the decline in woodland birds since 1970.

2 Suggest a possible reason for the decline in farmland birds since 1970.

3 The graph goes up to 2008. If further data had been published for 2010, what do you think they might show had happened to the populations of the different types of bird? Explain your answer.

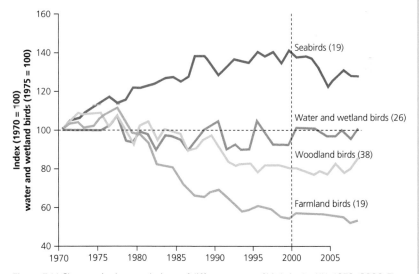

Figure 7.11 Changes in the populations of different types of birds in the UK, 1970–2008. The numbers in brackets indicate the numbers of species monitored.

85

► How can we get data about biodiversity in an environment?

Unless the environment studied is very small, the whole of it cannot be explored to find all the animals and plants living there. Plants are easier to find than animals, because they don't move around or hide, but even so it would be impossible to count all the plants in, for example, a woodland with an area of several square kilometres. The only way we can get an idea of numbers is to take a sample. A small area is studied in detail, and the numbers used to predict the population numbers in the environment as a whole. For example, if we wanted to study a snail species in an area of marshland that is 100 km², we might count all the snails in an area of 100 m × 100 m, made up of many smaller samples. The whole sample area is 10 000 m², which is 1/10 000 of the total marshland area. As an example, let us say there were 115 snails in the sample area, then:

sample area	= 1/10 000 of total area
number of snails in sample area	= 115
number of snails in total area	= 115 × 10 000
	= 1 150 000

For this number to be reasonably accurate, certain criteria need to be met:

▸ The sample area must be typical of the total area. Very small areas are more likely to be unusual in some way, and so the bigger the sample area is, the better.
▸ The method of sampling must not affect the results (for example, with some animals, but not snails, the presence of humans might scare them away).

Samples cannot be absolutely accurate, and scientists often use statistical analysis that takes account of sample size when drawing conclusions.

To sample a given area, scientists often use a piece of equipment called a quadrat. This is a frame of some sort with equal sides of a known length (Figure 7.12).

The quadrat is used many times to build up a bigger sample area. It should be put down at random, to avoid the experimenter introducing any sort of bias into the data collection.

Figure 7.12 Using a quadrat. In this case, the quadrat is 0.5 m × 0.5 m, giving an area of 0.25 m² for each quadrat.

 Specified practical

Investigation into the distribution and abundance of organisms

Counting daisies

Apparatus

> quadrat, 0.5 m × 0.5 m
> measuring tape

Procedure

1 Choose an area of grassland to sample, and measure the total area that you wish to study.
2 Place the quadrat 'randomly'. You might choose to drop it over your shoulder, without looking where it will land. Always shout a warning to make sure no-one is behind you, or they may get hit with the quadrat.
3 Count the number of daisy plants (not just the flowers) in your quadrat.
4 Repeat another nine times (that is, 10 times in total).
5 Use your data to calculate how many daisies are in the whole area of grassland.

Evaluating your experiment

Suggest any possible disadvantages of the method used to place the quadrat 'randomly'.

6 Explain why high biodiversity makes an ecosystem more stable.

7 Using part of a woodland to build houses on can reduce the biodiversity in the woodland. Why?

8 When sampling an area using a quadrat, why is it important to place the quadrats in a random way?

9 Scientists sampled an area of 1000 m² on a beach that had a total area of 1 km² (1 000 000 m²). They found 293 cockles. Estimate how many cockles there were on the whole beach.

10 Look at the environments in Figure 7.13. Suggest a reason why scientists would need to use a bigger sample area in woodland than in saltmarsh.

Figure 7.13 Woodland and saltmarsh environments.

▶ How can we find out about the distribution of organisms?

Sometimes, investigators don't want to just know what animals and plants are found in an environment, they also want to know something about their distribution. This can be investigated in a variety of ways, one of which is by doing a transect. A transect is a series of samples taken in a line. The line chosen for the samples usually lies along some sort of changing conditions – for example, when sampling a rocky shore, as in Figure 7.14, the transect might be lined up from the low tide level up to the high tide level.

Figure 7.14 Students taking quadrat samples along a transect line laid up and down a rocky shore.

Quadrats are laid down at regular intervals along the transect line and the animals and plants in each quadrat area are recorded. This allows any patterns of distribution to be detected (Figure 7.15).

Figure 7.15 The distribution of organisms along a transect can be plotted as a 'kite diagram'. The width of the line is a measure of how many organisms of that type were found in quadrats placed at intervals along the transect.

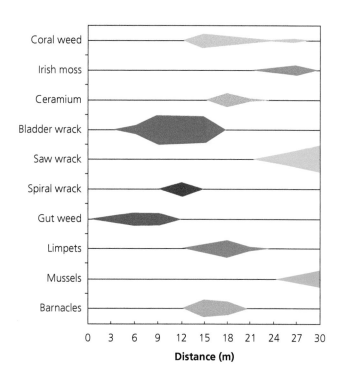

Specified practical

Investigation into the distribution and abundance of organisms

How does trampling affect plant distribution?

For this experiment, you will need to find a path through your school grounds or any area of grassland or woodland. It is best to test a path that has been worn away by people walking along the route, rather than a constructed path, though this is not essential. Some plants can survive being regularly trodden on, and these will be found on or near the path. The less resistant a plant species is to trampling, the further away from the path it will be, until you reach a distance where people rarely tread.

Apparatus

> quadrat, 0.5 m × 0.5 m
> 10 m string, marked every 0.5 m, or tape measure
> 2 skewers or similar to anchor string into ground
> plant identification book

Procedure

1 Lay the marked string across the path, extending a roughly equal distance on either side (Figure 7.16). If the path is wide, lay the transect line on just one side of the path.
2 Lay the quadrat down every 0.5 m along the transect, and record the species found and the numbers of each. Do your best to identify all the plants that you find.
3 Present your results in any suitable way.

Figure 7.16 Suggested transect line to study plant distribution across a path.

Analysing your results

1 Is there any pattern in your results?
2 Of the plants found, which would you say is the most resistant to trampling? Justify your answer using your data.
3 Are there any other factors, apart from trampling, that might affect plant distribution in the area you sampled?
4 Is there evidence of the influence of any other factor(s) in your results? Explain your answer.

How can we measure an animal population that moves around?

It is more difficult to measure animal populations in an area than plant populations, because animals move around. There is a danger of counting the same animal more than once, or of missing some that have just moved out of the sample area, but will return.

To determine the size of an animal population, the capture–recapture technique can be used. The technique works like this:

1 A number of individuals of a particular species are captured.
2 These animals are marked in some way so they can be distinguished from the rest of the population. They are then released back into the habitat.
3 Some time later, another sample of the species is captured.
4 The proportion of marked individuals in the second sample will be the same as the proportion of the initially marked individuals in the total population.

The total population can be estimated using the equation where:

$$N = \frac{MC}{R}$$

N = estimate of total population size
M = total number of animals captured and marked on the first visit
C = total number of animals captured on the second visit
R = number of animals captured on the first visit that were then recaptured on the second visit.

For the population estimate to be accurate, certain conditions must apply:

▶ Sufficient time must have elapsed between the taking of the two samples for the marked individuals to mix with the rest of the population.
▶ There must be no large-scale movement of animals into or out of the area in the time between the two samples.
▶ The marking technique must not affect the survival chances of the animal (for example, making it easier for a predator to see it).
▶ The marking technique must not affect the chances of recapture by making the marked individuals more 'noticeable' to the collector.

→ **Activity**

Capture–recapture technique calculation

Dave wanted to estimate the population of woodlice in his garden. He searched around and collected 100 woodlice. He marked each of them with a spot of white paint on its back, and then released them (Figure 7.17). A week later he went into his garden and collected another 100 woodlice. Four of those were marked woodlice that he had captured before.

Using the equation

$$N = \frac{MC}{R}$$

calculate the size of the woodlouse population in Dave's garden.

Figure 7.17 A marked woodlouse.

► Why can introducing a new species into a habitat cause problems?

If you go on holiday abroad, you are not allowed to bring back any plants (even seeds) or animals from the country that you visited. Similar restrictions apply in all countries. Have you ever wondered why?

Although biodiversity is good, there can be problems if you introduce an alien species (one not normally found in the area) into an environment. For example, the alien species may:

- ► have no predators in the area, and its population may grow out of control
- ► compete with an existing species, causing that species to die out in the area
- ► prey on existing species, reducing their number
- ► carry a disease to which it has immunity, but existing populations do not.

There are more than 3000 non-native species in the UK, more than in any other European country, so where have they all come from?

Some have arrived 'accidentally' – they may have arrived on board ships, amongst the cargo. Others may have been brought into the country by collectors or dealers and then escaped or been released (Figure 7.18).

Figure 7.18 The fallow deer is not native to Britain. It was brought to this country by the Normans (or possibly the Romans), probably for hunting.

Some species are deliberately brought in to control pest species. This is an example of biological control. Biological control involves using living organisms (often predators) instead of chemical pesticides. They are often used to control other alien species, which may have no natural predators in their new environment.

In the early days of biological control, the process sometimes went wrong, and the introduced predator itself caused a problem. One example occurred in the United States, where exotic thistles had been accidentally introduced and were reducing the

Figure 7.19 This beetle, *Rhinocyllus conicus*, was introduced into the USA to control alien species of thistle, but also ate the native thistles.

populations of the native thistles. A beetle, known to feed on the exotic thistles, was brought over in order to control them (Figure 7.19). However, when introduced into the area, the beetle also fed on native thistles. Certain local species of insect that fed only on the native thistles could no longer survive.

Scientists now understand the possible problems of introducing biological control agents and detailed research and extensive trials are now used before introducing any control species.

✔ Test yourself

11 Under what circumstances would you use a transect, rather than a random placement of quadrats?
12 When using the capture–recapture technique to estimate the size of a snail population, why would it not be a good idea to mark their shells with a large spot of bright fluorescent paint?
13 What is an 'alien' species?
14 If you are thinking of bringing in an alien species to act as a biological control, what information should you obtain about the alien species before introducing it, to avoid potential problems?

⬇ Chapter summary

- Living organisms show a range of sizes, features and complexities.
- Plants are broadly divided into flowering and non-flowering groups; animals are broadly divided into vertebrates and invertebrates.
- A scientific system for identification is needed to simplify and better organise the study of living things.
- Organisms with similar features and characteristics are classified into groups.
- Organisms are given scientific as opposed to 'common' names to avoid confusion that could be caused by local or national names.
- Organisms have morphological and behavioural adaptations that help them to survive in their environment.
- Individual organisms need resources from their environment – for example, food, water, light and minerals.
- The size of a population may be affected by competition for resources, as well as by predation, disease and pollution.

- Biodiversity is the variety of different species in an area, as well as the variation within those species.
- Biodiversity is important in maintaining the stability of an ecosystem.
- Biodiversity and endangered species can be protected by local and national measures.
- Quadrats can be used to investigate the abundance of species.
- Samples must be randomly placed, and sufficiently large, to be an accurate representation of the total area being investigated.
- The capture–recapture technique can be used to estimate population size for mobile organisms.
- The method of marking must follow certain criteria so as not to bias the results.
- The introduction of alien species can have harmful effects on local wildlife.
- Biological control agents can be effective, but there are issues surrounding their use that must be considered.

▶ Chapter review questions

1 Hamsters are common in many food webs.

Read the following statements about hamsters.

A They live in burrows underground.

B They come to the surface to feed.

C They have poor eyesight.

D They have sharp claws and strong front legs.

E They use a lot of the energy in their food as heat.

a) Which of the statements (A–E) suggests that hamsters:

  i) may rely on their sense of smell to find food *[1]*

  ii) have a high rate of respiration *[1]*

  iii) are adapted to dig burrows? *[1]*

Scientists investigated the proportion of a population of hamsters that were above ground during a period of 24 hours. The results are shown in the graph.

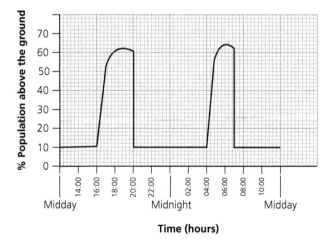

b) Use the graph to answer the following questions.

  i) Between which times during the 24-hour period were more than 10% of the population above ground? *[2]*

  ii) Foxes and owls are active mostly at night. What is the evidence that hamsters try to avoid being caught by foxes and owls? *[1]*

(from WJEC Paper B1(F), Summer 2013, question 2)

2 Rosebay willowherb, *Epilobium angustifolium*, is a plant that produces wind-dispersed seeds.

The survival of this plant in its natural habitat was studied by counting the numbers of:

> seeds found on the ground
> seedlings
> fully grown plants.

The counts were completed every 2 metres away from the parent population. All counts were taken in the direction of the prevailing wind (direction in which the wind mainly blows).

The results are shown in the table.

Distance from parent population, in m	Seeds, per m²	Seedlings, per m²	Fully grown plants, per m²
2	22	20	0
4	30	25	0
6	31	30	0
8	28	25	1
10	25	20	2
12	18	15	3
14	9	9	5
16	8	5	5
18	4	3	3

a) Name:

  i) the technique you would use to obtain the data shown in the table *[1]*

  ii) two items of apparatus used to make the necessary measurements for this technique. *[2]*

b) Calculate the percentage of seeds that survived to produce fully grown plants at 10 m from the parent plants. *[2]*

c) Explain why no fully grown plants are found within 6 m of the parent population. *[2]*

(from WJEC Paper B2(H), Summer 2015, question 9)

8 Cell division and stem cells

Specification coverage

This chapter covers the GCSE Biology specification section 2.2 Cell division and stem cells and GCSE Science (Double Award) specification section 4.2 Cell division and stem cells.

It covers the processes by which cells divide to cause growth (by mitosis) and to provide cells for sexual reproduction (by meiosis). There is consideration of how uncontrolled mitosis can result in cancer. The use of stem cells in replacing damaged tissue is discussed.

▶ Why is cell division important?

A human body has about 50–100 trillion cells in it (depending on size). Yet every human being started out as a single cell inside its mother's body. The cells in the body are constantly being replaced. For example, around 2 million red blood cells are formed (and another 2 million destroyed) every second! All new cells are formed by division of existing cells, as we have already seen in the cell theory (Chapter 1). Bacteria actually reproduce themselves by cell division, because they are only one cell big anyway.

In order to function properly, each new cell needs a set of genes. These genes are contained in chromosomes, which are duplicated and passed on in cell division.

▶ What is a chromosome?

Chromosomes are found in the nucleus of every cell. You cannot see them normally, because they are very long, incredibly thin strands of DNA. DNA is an important molecule, which you will learn more about in Chapter 9. Just before cell division, the DNA coils up tightly and this allows the chromosomes to be visible under a microscope. The appearance of chromosomes in an onion root cell is shown in Figure 8.1.

Sections of DNA that control features of the organism are called genes. A chromosome is basically a long line of genes. The number of chromosomes in a cell varies between different species – human body cells have 46. As we shall see later, you receive 23 chromosomes from your mother and 23 from your father, so the 46 chromosomes are actually made up of 23 pairs. The pairs are not identical, but do look the same and they have the same genes. They are not identical because the form of the gene (called the allele) can vary. For example, the gene that can cause the disease cystic fibrosis is found

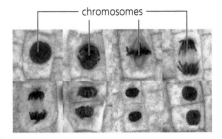

Figure 8.1 Chromosomes in onion root cells.

on chromosome number 7. There are two number 7 chromosomes. One or both may have the 'healthy' form of the gene, and one or both may have the form that causes cystic fibrosis (Figure 8.2).

The full set of human chromosomes, arranged in their pairs, is shown in Figure 8.3. These are the chromosomes of a male, and males have one 'pair' of chromosomes that do not look the same (the X and Y chromosomes). Females have two X chromosomes.

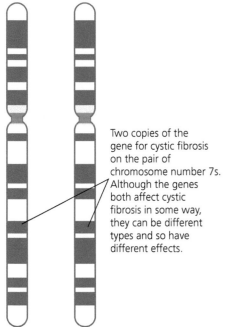

Two copies of the gene for cystic fibrosis on the pair of chromosome number 7s. Although the genes both affect cystic fibrosis in some way, they can be different types and so have different effects.

Figure 8.2 Chromosome pair number 7, showing the position of the cystic fibrosis gene.

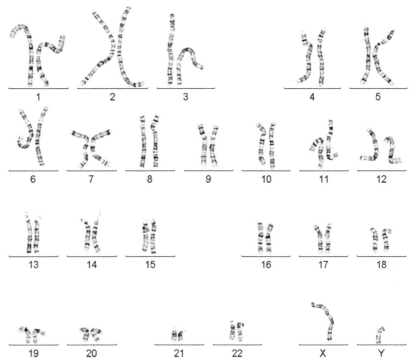

Figure 8.3 The chromosomes of a human male, arranged in their pairs.

How do new cells form?

In multicellular organisms, cell division results in growth, and repair and replacement of old or damaged cells and tissues. The type of cell division that occurs in these processes is called mitosis, where one cell (the 'mother' cell) divides to form two new ('daughter') cells. The daughter cells are genetically identical to the mother cell, because the chromosomes replicate (copy) themselves and one copy is passed into each of the two new cells (Figure 8.4). The number of chromosomes is the same in the daughter cells as in the mother cell.

Mitosis is the normal type of cell division, but there is another type. This is called meiosis, and it only occurs when sex cells (gametes) are formed. In humans, all the body cells have 46 chromosomes, and mitosis produces new cells with 46 chromosomes. When forming gametes, though, it is important that the sperm and egg cells do not have 46 chromosomes. If they did, when the sperm fertilised the egg, the resulting zygote would have 92 chromosomes, and would not produce a normal human being.

In meiosis, although the DNA and chromosomes duplicate as in mitosis, four new cells are formed instead of two, and each cell receives just half of a full set of chromosomes. In humans, therefore, the sperm and egg cells each have 23 chromosomes, so that a new

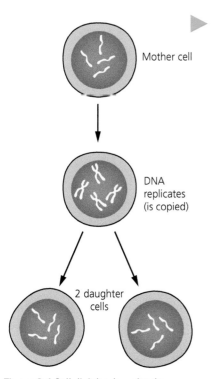

Mother cell

DNA replicates (is copied)

2 daughter cells

Figure 8.4 Cell division by mitosis.

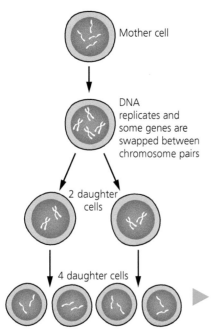

Figure 8.5 Cell division by meiosis.

zygote will have 46 chromosomes as it should. The chromosomes come in pairs that are not identical, because during the process of meiosis the chromosome pairs swap genes with each other as they are lining up, and the gametes get one chromosome from each pair. The new cells in meiosis, unlike mitosis, are therefore not genetically identical. Meiosis is shown in Figure 8.5.

Table 8.1 shows the differences between the two types of cell division.

Table 8.1 Comparing mitosis and meiosis.

Mitosis	Meiosis
Occurs in all body cells *except* those forming gametes	Occurs only in gamete-forming cells
Daughter cells are genetically identical	Daughter cells are genetically different
Two daughter cells are formed	Four daughter cells are formed
Daughter cells have a full set of chromosomes	Daughter cells have a half set of chromosomes

What happens if mitosis goes wrong?

Cells normally go through a cycle of cell division. A cell divides, and the two new cells then spend time growing and developing into mature cells, before dividing again. How long this takes depends upon the type of cell. The speed of the cycle is regulated by special genes, but occasionally a fault develops that results in cell division happening too frequently, without giving time for the cells to mature. This uncontrolled growth leads to cancers. The cells grow into a tumour, which damages the organ or tissue it is in (Figure 8.6). If the cells reach the blood system they can be taken around the body, still dividing, and they can set up new tumours wherever they may lodge in other organs.

Figure 8.6 This cancerous tumour was removed from a woman's stomach. It weighed 38 kg.

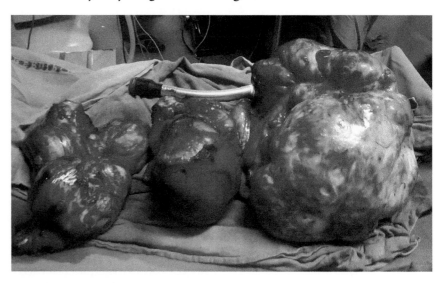

What are stem cells?

In the last few years there has been a lot of controversy about the use of stem cells. Stem cells are undifferentiated cells, but what does that mean?

When a plant or animal embryo is formed and starts to grow, the cells all appear the same. Eventually, the cells start to

Practical

Observing cell division

In plants, mitosis is concentrated in special growing points in the stems, buds and roots. When the cells are stained with ethano-orcein stain, the chromosomes in dividing cells can be seen. To prepare for this experiment observing cell division in root cells, onions (or garlic cloves) have to be kept with their bases just in contact with water in a beaker, in the dark, for several days (Figure 8.7). The roots need to be 2–3 cm long.

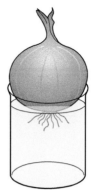

Figure 8.7 Growing onion roots in a beaker of water.

Apparatus and chemicals

> onion or garlic roots
> ethano-orcein stain
> 1 mol/dm^3 hydrochloric acid
> watch glass with cover
> microscope slide
> cover slip
> scalpel
> Bunsen burner

> tongs
> dropping pipette
> filter paper
> forceps
> mounted needle
> water bath at 55 °C
> 2 × 100 cm^3 beakers
> microscope

Procedure

1 Cut off about 5 mm of root tips, and place them in a beaker of cold water to wash them for 4–5 minutes and dry on filter paper.
2 Heat 10–25 cm^3 of hydrochloric acid in one of the beakers in the water bath.
3 Transfer the root tips to the hot hydrochloric acid and leave for 5 minutes.
4 Wash the root tips in water again for 4–5 minutes, and dry as before.
5 Place one of the root tips on a microscope slide. Cut the root tip to leave just the terminal 1 mm. Discard the rest.
6 Add a drop of ethano-orcein stain, and leave for 2 minutes.
7 Gently 'mash' the root tip with a mounted needle.
8 Cover the tip with a cover slip.
9 Gently squash the tip by tapping it with the blunt end of a pencil or mounted needle about 20 times. This is best done by dropping the pencil vertically onto the cover slip from a height of about 5 cm.
10 Look at the slide under the microscope and try to find the growing area, where chromosomes will be seen in the cells. See the example in Figure 8.8.
11 Draw two or three dividing cells.

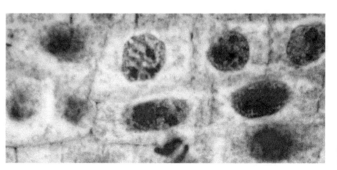

Figure 8.8 Stained onion root cells showing cell division.

differentiate – to become specialised in some way, as a liver cell, a nerve cell or an epidermal cell, for example. Once they have differentiated, if they then divide, they can only form similar cells to themselves. A liver cell can never become a nerve cell. The undifferentiated cells in the embryo, though – the stem cells – can become any cell at all. Scientists can take stem cells from an embryo and grow them into types of cells that can be used to repair or replace damaged tissue. This may eventually allow treatment of diseases and conditions such as cancer, type 1 diabetes, brain damage, spinal cord injury and so on. The only trouble is that the embryo, which could potentially grow into a human being, is destroyed in the process. The embryos currently used for research are those that weren't used during *in vitro* fertilisation (IVF – treatment to help infertile couples have a baby), but there is a possibility that embryos could be created purely to supply stem cells, and some people feel that is wrong.

There are alternatives, however. Stem cells can be found in adults (for example, in the bone marrow inside bones). These cells are in mature tissues but, unusually, have not lost the ability to differentiate into different cells. Stem cells can also be collected from the blood from the umbilical cord at birth.

The growing areas of plants (the root and shoot tips), called meristems, also produce cells that can differentiate into other cells (but only cells of that particular plant, so they have no medical use).

→ | Activity

How should we use stem cells, if at all?

Figure 8.9 shows some opinions that are held by different people about stem cells. Research stem cells, then pick one of the opinions (that you agree with) and write a letter to a newspaper, explaining your opinion and backing it up with evidence.

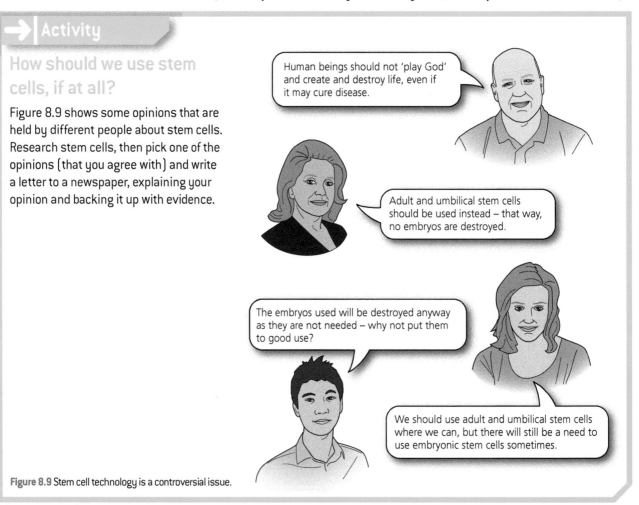

Human beings should not 'play God' and create and destroy life, even if it may cure disease.

Adult and umbilical stem cells should be used instead – that way, no embryos are destroyed.

The embryos used will be destroyed anyway as they are not needed – why not put them to good use?

We should use adult and umbilical stem cells where we can, but there will still be a need to use embryonic stem cells sometimes.

Figure 8.9 Stem cell technology is a controversial issue.

✔ **Test yourself**

1 How many chromosomes does a human body cell have?
2 Why are the daughter cells formed by mitosis genetically identical?
3 Cats have 38 chromosomes, dogs have 78 and wheat has 42. How many chromosomes would you expect to find in:
   a) an egg cell of a dog
   b) a kidney cell of a cat
   c) a pollen cell of wheat?
4 Why would meiosis not work as the 'normal' method of cell division in the body?
5 Why are adult stem cells less useful than embryonic stem cells?

▼ **Chapter summary**

- Cell division by mitosis enables an organism to grow, and to replace and repair cells.
- In mitosis, the number of chromosomes remains constant and the daughter cells are genetically identical to the mother cell.
- Sex cells (gametes) are formed by a different form of cell division called meiosis.
- In meiosis, the number of chromosomes is halved and the daughter cells are not genetically identical.
- Mitosis produces two daughter cells, while meiosis produces four.

- If mitosis is uncontrolled, it can result in cancer.
- In mature tissues, the cells have usually lost the ability to differentiate into different forms.
- In both plants and animals, certain cells, called stem cells, are capable of differentiating into different forms of cell.
- Human stem cells have the potential to replace damaged tissue and could be the basis of treatment for a variety of diseases and conditions.
- Human stem cells can be obtained from embryos and from adult tissues.

▶ Chapter review questions

1. Barack Obama, the President of the United States of America from 2008 until 2016, supports research into the use of embryonic stem cells. However, Newt Gingrich, who was hoping to become President, said in February 2012, that he would 'ban embryonic stem cell research if he became President'.

   Suggest why some people support embryonic stem cell research, whereas others do not. [2]

   (from WJEC Paper B1(F), Summer 2013, question 2)

2. In December 2010, a dog named Boris was treated for severe arthritis of the hip joints in a veterinary clinic in West Michigan, USA. Some of the stages in the treatment are shown below.

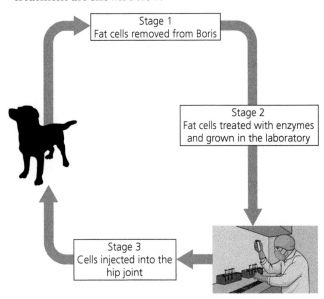

   Three months after treatment, Boris was examined at the veterinary centre. His hips were found to have greatly improved and X-rays of the hip joints showed evidence of repair of the joint tissues.

   a) State what type of cells are injected in stage 3 in the diagram above. [1]

   b) State one advantage of this method of treatment over the use of embryonic stem cells. [1]

   (from WJEC Paper B2(H), Summer 2013, question 1)

3. The diagram below shows a human cell dividing by mitosis to form two new cells.

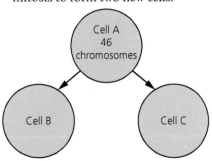

   a) Complete the diagram by writing the number of chromosomes in each of the two new cells. [1]

   b) Mitosis enables organisms to grow. State one *other* function of mitosis. [1]

   c) Copy and complete the table, which compares mitosis and meiosis. [2]

	Mitosis	Meiosis
Number of new cells from each division	two	
Genes in new cells compared to original cell		different

   d) What is the scientific term for the sex cells (eggs and sperm) that are produced by meiosis? [1]

   (from WJEC Paper B2(F), Summer 2014, question 3)

4. Read the information about stem cells and answer the questions that follow.

   To be called a stem cell, a cell needs two properties:

   a) The cell can go through numerous cycles of cell division, while remaining undifferentiated.

   b) The cell has the ability to develop into several or many different types of cell.

   Stem cells are either **pluripotent** cells, which can develop into almost any type of cell, or **multipotent**, which means they can develop into several different types, but only those that are closely related to each other in what is called a 'family' of cells.

   Stem cells can be found and extracted from early stage embryos and these cells are pluripotent. Stem cells can also be found in the bone marrow of children and adults. These are known as adult stem cells and they are multipotent. If a patient's own stem cells are used to treat a condition, then the risk of rejection is virtually non-existent.

   At the moment, investigations are being carried out to develop stem cell treatments for many diseases, including diabetes, Parkinson's disease, Alzheimer's disease and stroke.

   a) What type of cell division would happen in embryonic stem cells? [1]

   b) What does the term 'undifferentiated' mean? [1]

   c) From information in the passage, state one advantage and one disadvantage of using adult stem cells rather than embryonic stem cells. [2]

   d) If a treatment for diabetes were developed, which organ would need to be treated? (Note – you will need information from Chapter 11 to answer this question.) [1]

   e) Give a reason why some people object to the use of embryonic stem cells. [1]

9 DNA and inheritance

 Specification coverage

This chapter covers the GCSE Biology specification section 2.3 DNA and inheritance and GCSE Science (Double Award) specification section 4.3 DNA and inheritance.

It covers the structure of DNA and how it acts as a code for the production of proteins and therefore produces the differences seen between different individuals. The application of genetic profiling as an application for looking at differences between individuals is studied. The mechanisms of inheritance are also covered, including the use of Punnett squares.

▶ How does the nucleus control the cell?

We saw in Chapter 1 that all the chemical activities in cells are controlled by enzymes. Enzymes are proteins, and the 'instructions' for making the enzymes (and other proteins) are stored in the nucleus, in the form of a chemical called DNA (deoxyribonucleic acid).

DNA is the chemical that makes up your genes, controlling the structure and function of your body by controlling the production of proteins. Besides enzymes, many other important molecules in the body are made of protein – including hormones and antibodies. Proteins are also the main constituents of all the body's tissues (for example, muscle).

All proteins are made up of long chains of molecules called amino acids. The chains are coiled and folded to give every protein a specific shape, and we have already seen how important this is in enzymes.

DNA contains a sort of chemical code, which tells the cell which amino acids to assemble in order to make a protein. DNA is a very strange molecule, because it can make copies of itself. Whenever a new cell is made, it has to have a set of genes, so the DNA duplicates itself and one set of genes is passed into each new cell.

DNA is made up of two long chains of alternating sugar and phosphate molecules connected by pairs of bases. This ladder-like structure is twisted to form a 'double helix' (a helix is a type of spiral). There are four bases in DNA: adenine (A) joins on to thymine (T), and guanine (G) joins on to cytosine (C). The order of these bases along the sugar–phosphate backbone varies in different molecules of DNA. This sequence of bases forms the instructions, in a form of code, for the manufacture of proteins. It determines which amino acids are used to make a given protein, and in what order. The structure of DNA is shown in Figure 9.1.

The 'code' consists of **triplets** (groups of three) of bases along the DNA. Each triplet codes for an individual amino acid in the protein.

'Backbone' chains of alternating sugar and phosphate units, twisted into a double helix

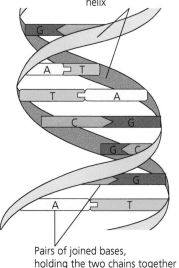

Pairs of joined bases, holding the two chains together

Figure 9.1 The structure of DNA

What is a gene?

In the nucleus of a cell, the long DNA molecules are coiled up into structures called **chromosomes**. As we have seen, DNA is the raw material of genes – a **gene** is a short length of DNA that codes for one protein. This is summarised in Figure 9.2.

In Figure 9.2 the coloured bands seen running across the centre of the DNA molecule are the pairs of bases referred to in the last section. The sequences of these bases in a DNA molecule form the 'code' that decides what proteins the cell makes.

Figure 9.2 The structure of a gene in relation to DNA and chromosomes.

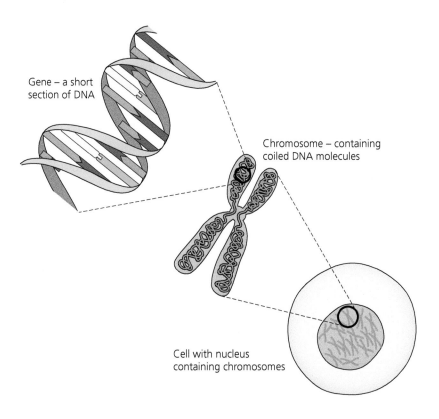

Gene – a short section of DNA

Chromosome – containing coiled DNA molecules

Cell with nucleus containing chromosomes

Scientists can look at the bases in a DNA molecule and see to what extent different DNA samples are similar. The analysis of the DNA produces a genetic profile. Everyone's genetic profile is slightly different, and this has led to police scientists being able to identify the person who may have left some DNA at a crime scene ('genetic fingerprinting'). It has also been used to identify the father of a child, where this is not certain, and to establish how closely related two species are and so help classify them.

Genetic profiling consists of several steps:

1 A sample of cells is collected – for example, from blood, hair, semen or skin. The cells are broken up and the DNA extracted.
2 The DNA is 'cut up' by enzymes, so that it ends up in fragments of different sizes.
3 The fragments are then separated. A pattern develops, which is the 'genetic profile'.

Genetic terminology

The study of the way that genes are inherited is called **genetics**. The next section looks at how inheritance works. There are some special genetic terms you will need to know and be able to use, in order to understand and explain genetics.

Key terms

Gene A length of DNA that codes for one protein.

Allele A variety of a gene.

Chromosome A length of DNA that contains many genes, found in the nucleus and visible during cell division.

Genotype The genetic make-up of an individual (for example, **BB**, **Bb**, **bb**).

Phenotype The description of the way the genotype 'shows itself' (for example, blue eyes, curly hair, red flowers, and so on.)

Dominant The allele that shows in the phenotype whenever it is present (shown by a capital letter – for example, **B**).

Recessive The allele that is hidden when a dominant allele is present (shown by a lower case letter – for example, **b**).

F1/F2 Short for first generation (F1) and second generation (F2) in a genetic cross.

Homozygous/homozygote A homozygote contains two identical alleles for the gene concerned – it is homozygous.

Heterozygous/heterozygote A heterozygote contains two different alleles for the gene concerned – it is heterozygous.

Selfing A technique by which pollen from a plant is used to fertilise ovules in flowers of the same plant.

▶ ## Gregor Mendel and the inheritance of a single gene

The first person to use the term 'gene' was the Danish scientist Wilhelm Ludwig Johannsen in 1902, but the idea of the gene had been established 40 years earlier by an Austrian monk, **Gregor Mendel**, who is now regarded as one of the most famous scientists of all time (Figure 9.3).

When Mendel did his experiments on pea plants, scientists knew that characteristics were inherited from parents, but thought that the inherited character was a sort of 'blend' of the parents' characters. Mendel used 'tall' and 'dwarf' pea plants that were 'pure breeding – that is, they always produced the same type of offspring. Both types

Figure 9.3 Gregor Mendel, the 'father' of genetics.

varied in size a bit, but there was a clear difference in the range of heights of the two categories. He crossed tall plants with dwarf ones by careful pollination, but the offspring (the F1 generation) were not all 'medium sized' plants, as expected. In fact, all the new plants were tall (Figure 9.4).

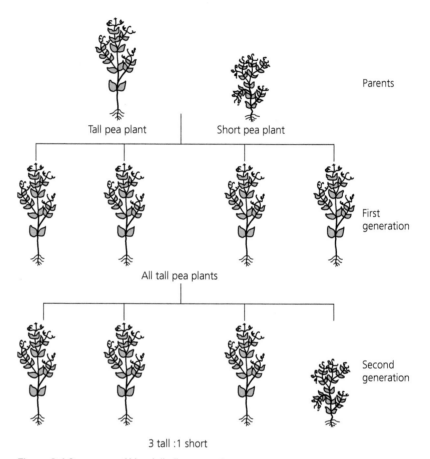

Figure 9.4 Summary of Mendel's first experiments.

When Mendel crossed two of these tall F1 plants together, he got offspring (the F2 generation) that were mostly tall plants, but about a quarter were dwarf. He now knew that whatever caused the 'dwarf' feature *had* been inherited in the first cross, but for some reason had been masked. The real genius of Mendel was how he investigated and explained what had happened.

Mendel knew that, if the dwarf character had been masked, another (tall) factor must be masking it. Therefore, the pea plants had two factors that controlled height, and the tall factor seemed to be in some way stronger than the dwarf one. We now call these different factors alleles, and we refer to 'stronger' ones as dominant and 'weaker' ones as recessive.

Mendel worked out how this pattern of inheritance had come about. It can be shown by a diagram, in which the dominant allele (tall) is shown by the symbol **T**, and the recessive dwarf allele is shown by the symbol **t** (Figure 9.5). Geneticists always show dominant alleles in capital letters and recessive ones in lower case. The genetic make-up of an organism is called its genotype and what it 'looks' like is called its phenotype.

Figure 9.5 A genetic diagram explaining Mendel's experiments.

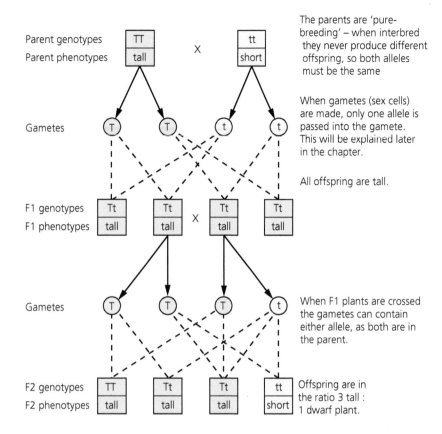

Parent genotypes TT X tt
Parent phenotypes tall short

The parents are 'pure-breeding' – when interbred they never produce different offspring, so both alleles must be the same

Gametes T T t t

When gametes (sex cells) are made, only one allele is passed into the gamete. This will be explained later in the chapter.

All offspring are tall.

F1 genotypes Tt Tt X Tt Tt
F1 phenotypes tall tall tall tall

Gametes T T T t

When F1 plants are crossed the gametes can contain either allele, as both are in the parent.

F2 genotypes TT Tt Tt tt
F2 phenotypes tall tall tall short

Offspring are in the ratio 3 tall : 1 dwarf plant.

In crosses like the second one in Figure 9.5 the use of arrows can be a bit confusing. Biologists tend to put the gametes in a table called a **Punnett square**, as shown in the activity on the next page.

Mendel tried his experiments with a number of characteristics apart from height. In each case, he found that there was a dominant and a recessive characteristic. In the initial crosses, all the F1 offspring showed the dominant phenotype. When two F1 plants were crossed, the ratio between dominant and recessive phenotypes in the F2 generation was around 3 : 1. Note that the ratio means that the dominant form is three times more likely than the recessive form, but does not mean that there will always be *exactly* three times as many plants showing the dominant phenotype than there are showing the recessive phenotype.

Mendel's conclusions were:

▷ Characteristics are controlled by a pair of factors (alleles), which may be the same as each other or different.
▷ One allele is dominant, the other is recessive. If both alleles are present, only the dominant one can be seen in the phenotype.
▷ Only one allele from each pair is passed into each gamete.
▷ If two pure-breeding parents (one for each allele) are crossed, the F1 offspring all show the dominant characteristic.
▷ If two of the F1 offspring are crossed, the new F2 generation has approximately three individuals showing the dominant characteristic to every one showing the recessive characteristic.

Mendel was able to make his discoveries because inheritance in peas is relatively simple, and many characteristics are controlled by a single pair of alleles.

In humans this is rare; most of our features are controlled by a combination of genes. It is known, for instance, that in humans more than 400 genes influence height, although the exact number is not known as scientists are still learning about all the effects of human genes.

✔ Test yourself

6 Mendel's scheme predicted a 3 : 1 ratio in F2 plants. His results all showed a roughly 3 : 1 ratio, but never an exact 3 : 1 ratio. Does this matter?

7 In some people, the lower part of the ear has a 'lobe' and in other people it is attached to the head. It is thought that lobes are caused by a dominant allele, **E**. Attached ears are due to a recessive allele, **e**. If someone has a genotype of **Ee**, would they have ear lobes?

8 In an animal, black coat colour (allele **B**) is dominant to white (allele **b**). Use a Punnett square to work out the expected ratio of black and white animals from a cross between a heterozygote (**Bb**) and a homozygous recessive animal (**bb**).

9 In peas, a round seed coat (allele **R**) is dominant to a wrinkled seed coat (allele **r**). A gardener crosses a pea plant with round seeds with another with wrinkled seeds. After many such crosses, he has never got offspring plants with wrinkled seeds. What does that tell him about the genotype of his round seed plants?

→ Activity

Using a Punnett square

Question

In the 'Four o'clock' plant (*Mirabilis jalapa*) red flower colour is dominant to white (Figure 9.6).

Figure 9.6 Red and white flower variants in *Mirabilis jalapa*.

When a pure-breeding red-flowered plant was crossed with a pure-breeding white-flowered plant, all the offspring were red. One of these red-flowered plants was then 'selfed' (self-pollinated). Predict the outcome of this self-pollination.

Answer

Let us call the dominant red allele **R** and the recessive white allele **r**.

The original red-flowered plants therefore had the genotype **RR**, and the white ones had the genotype **rr**.

The cross between these plants could only create the genotype **Rr**, so all of the F1 red plants must have been **Rr**.

When these plants were selfed, therefore, the cross would have been

Rr × Rr.

Before doing the Punnett square, write down the genotypes and phenotypes of the parent plant and the gametes.

	Male	Female
Parent genotype	Rr	Rr
Parent phenotype	Red	Red
Gametes	R or r	R or r

Now do the Punnett square.

		Male gametes (in pollen)	
		R	r
Female gametes (in egg cell)	R	RR	Rr
	r	Rr	rr

Now identify the genotypes of the offspring. **RR** and **Rr** are red; **rr** is white. You can see that this cross would produce red-flowered and white-flowered plants in the ratio 3 red : 1 white.

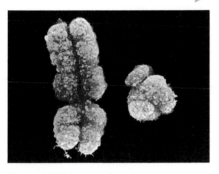

Figure 9.7 This scanning electron micrograph shows an X chromosome and a Y chromosome, from a human male. Females have two X chromosomes.

Figure 9.8 Inheritance of sex in humans.

► How is sex determined?

The last chapter described how chromosomes are arranged in pairs. Humans have 23 pairs, and in general, the members of each pair look the same, but there is one exception. Pair number 23 is different in males and females. In females the two chromosomes do look the same, but in males they are different from each other. Because of their shape, the larger chromosome is called the X chromosome and the shorter one is called the Y chromosome. Males have one X and one Y chromosome (Figure 9.7), while females have two X chromosomes.

This pair of chromosomes determines whether the individual is male or female. When egg cells are formed in a female's ovary, they all contain an X chromosome (because there is no alternative), but when sperm form in a male's testis, half the sperm cells have an X chromosome and half have a Y chromosome. At fertilisation, when an egg cell fuses with a sperm cell carrying an X chromosome, a female embryo will develop, while an egg fusing with a sperm carrying a Y chromosome will develop into a male embryo. So when a woman becomes pregnant, there is a 50% chance of having either a boy or a girl, as shown in Figure 9.8, because sperm and egg

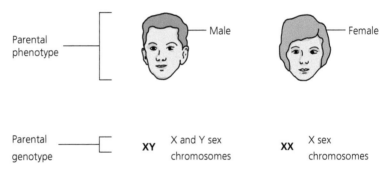

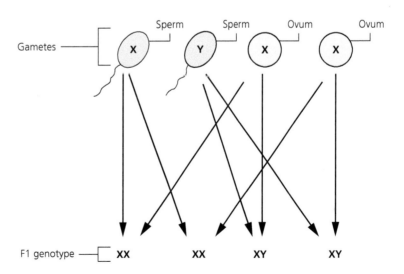

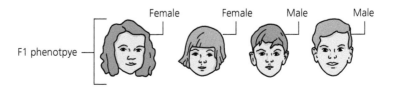

cells combine at random, and roughly half of the sperm carry an X chromosome and half carry a Y chromosome.

▶ Is it a good thing to change genes artificially?

Scientists can now extract genes from one organism and put them into another, and can also 'swap' one gene for another. The introduction of genes into food plants is becoming more common and is known as **genetic modification**. In the 1980s, the first commercial genetically modified (GM) crop was developed that was resistant to insects and pests. It was the potato, and it was modified so that it made its own built-in insecticide. The insecticide was an insect poison normally produced by a type of bacterium that lives in the soil. The gene for the poison's production was transferred to potato plants, which then made the plant resistant to insect pests.

Resistance to herbicide is now common in genetically modified crops. By 2007, more than 50% of soya harvested across the world was genetically modified. In 2010, the European Commission approved a measure to allow different countries to choose for themselves whether or not to develop genetically modified crops.

Weeds compete with crops if left unchecked. For a great many years farmers have attempted to get rid of weeds by using chemicals called herbicides. However, selective herbicides that kill only weeds and not the crop plants are difficult to produce. A herbicide-resistance gene can be taken from a bacterium that normally grows in soil and transferred to a plant such as soya.

Unfortunately, there have been some potential problems with this technology, to which some people have objected. For example, there is a possibility of herbicide-resistant plants escaping into the environment and flourishing. How could they be destroyed if herbicides cannot kill them? The answer is to ensure that the plants are sterile and can only reproduce asexually. Another unwanted side effect found in herbicide-resistant soya was that the stems of many plants split in hot climates and could not support the plant.

The advantages of herbicide-resistant and insect-resistant plants are that far fewer chemicals need to be introduced into the environment to kill insects and weeds. Theoretically, high yields from crops can be maintained without affecting the environment. However, GM crops are new technology and more scientifically valid trials need to be carried out to decide if they are beneficial. There seem to be both advantages and disadvantages to GM crops.

The case for GM:

▸ Crops could be tailor-made to suit the varied farming conditions found throughout the world. In this way they could provide more nutritional value and a higher income.
▸ Energy-producing crops could save natural resources and so conserve the environment.

The case against GM:

▸ GM crops could reduce the developed countries' reliance on crops from developing countries. This could result in loss of trade and severe economic damage for the developing countries.

▶ It is difficult to stop the pollen from GM crops grown in fields from pollinating other nearby crops. People who do not want to grow GM crops (such as organic farmers) may still end up with modified genes in their crop.

▶ The companies that develop a GM crop have the 'patent' on it, meaning that they are the only ones that can distribute it. This means that they will control the price, which may be too expensive for poorer countries.

These issues raise important political, ethical and economic questions that are not unique to modern biotechnology. They must be resolved at government and international level to maximise the benefits from gene technology.

 Test yourself

10 Is the sex of a human baby decided by the egg cell or the sperm cell that formed it? Explain your answer.
11 Why is it reasonable to expect that the ratio of boy babies to girl babies in a large human population will be around 1 : 1 (that is, 50% boys and 50% girls)?
12 If a plant is described as a GM crop, what does that mean?
13 Why is it difficult to stop GM crops pollinating other non-GM varieties growing in nearby fields?

Chapter summary

- DNA consists of two long chains of alternating sugar and phosphate molecules connected by bases; the chains are twisted to form a double helix.
- There are four types of bases – A, T, C and G – and the order of bases along the DNA molecule forms a code for making proteins; the code determines the order in which different amino acids are linked together to form different proteins.
- In DNA, the base adenine pairs with thymine, and cytosine with guanine.
- A sequence of three bases (a triplet) determines one amino acid to be added to a protein.
- Genetic profiling involves cutting the DNA into short pieces using specific enzymes. The fragments are then separated into bands according to their size, by gel electrophoresis.
- The pattern of the bands produced can be compared to show the similarities and differences between two DNA samples – for instance, in criminal cases, paternity cases and in comparisons between species for classification purposes.
- DNA profiling can be used to identify the presence of certain genes, which may be associated with a particular disease.
- Genes are sections of DNA molecules that determine inherited characteristics.

- Genes have different forms, called alleles. In a pair of chromosomes, there may be two different alleles of a particular gene, or the two alleles may be the same.
- The following terms are used in genetics: gamete, chromosome, gene, allele, dominant, recessive, homozygous, heterozygous, genotype, phenotype, F1, F2, selfing.
- Punnett squares can be used to show the inheritance of single genes.
- The cross **Aa** × **Aa** gives a 3 : 1 ratio of dominant : recessive phenotypes among the offspring.
- The cross **Aa** × **aa** gives a 1 : 1 ratio of dominant : recessive phenotypes among the offspring.
- Most phenotypic features are the result of multiple genes rather than single gene inheritance.
- Sex determination in humans is the result of the composition of the pair of sex chromosomes; females have two X chromosomes, males have one X and one Y chromosome.
- The sex chromosomes separate in the gametes, and combine randomly at fertilisation.
- The artificial transfer of genes from one organism to another is known as genetic modification.
- Genetic modification has potential advantages and disadvantages.

▶ Chapter review questions

1. In tigers the normal fur colour is mainly orange with black stripes. This orange colour is caused by a dominant allele **R**.

Over a number of years, a pair of heterozygous orange tigers named Sashi and Ravi produced 13 cubs in a Zoo in India. Three of these cubs were white (that is, white with black stripes).

a) State the genotypes of Sashi and Ravi. *[1]*

b) Draw a Punnett square to show how the white cubs were produced. *[2]*

c) Draw a Punnett square to show how an orange tiger mated with a white tiger could produce offspring that were all orange. *[2]*

(from WJEC Paper B1(H), January 2011, question 1)

2. The following front page headline appeared in the Western Mail newspaper in January 2008.

> **Outrage as one in ten of us is on the DNA database**
>
> Almost one in ten people in Wales are on the national DNA database. Many of the 264 420 on the database have never been charged with any criminal offence but their DNA sample is kept for life.

a) Suggest one reason why some people, who have never been charged with an offence, object to their DNA samples being kept on record. *[1]*

b) Suggest one advantage of the police keeping a DNA database. *[1]*

(from WJEC Paper B1(H), Summer 2009, question 1)

3. In sheep, white colour (**D**) is dominant to black colour (**d**). A white ewe (female) was crossed with a black ram (male). All the F1 offspring were white.

a) What are the genotypes of:

   i) the white ewe   ii) the black ram? *[1]*

b) Draw a Punnett square to show the potential genotypes of offspring of a mating between the white ewe and the black ram. *[2]*

c) Draw a Punnett square to show the potential genotypes of F2 offspring if two of the F1 offspring are mated together. *[2]*

d) In the cross described in (c), what would be the expected ratio of the genotypes homozygous white : heterozygous white : homozygous black? *[1]*

(from WJEC Paper B1(H), Summer 2009, question 2)

4. Gregor Mendel is the person credited with the discovery of the principles of genetics. He worked with pea plants, and one of the features he looked at was pod colour. In peas, a green pod is dominant to a yellow pod. In one cross, Mendel crossed two heterozygous pea plants, both with green pods, but each carrying what he called a yellow 'factor'. He repeated this cross 580 times, and the plants produced 428 green pods and 152 yellow pods.

a) Mendel described the cause of the pod colour as a 'factor'. What do we call these 'factors' today? *[1]*

b) The green 'factor' can be represented by the letter G and the yellow 'factor' by the letter g. Draw a punnet square of the cross that Mendel did (Gg × Gg) and state the expected ratio of green pods to yellow pods. *[3]*

c) What was the actual ratio of green pods : yellow pods in Mendel's results? *[2]*

d) The actual ratio does not match the expected ratio. Why does this not mean that the expected ratio is wrong? *[1]*

e) The peas used in this cross were **heterozygous**. What does that mean? *[1]*

f) Of the 428 green pods, how many (approximately) would you expect to be **homozygous**? *[2]*

g) State one reason why we can have confidence in Mendel's results. *[1]*

10 Variation and evolution

▶ Am I unique?

Yes, you are. No human being is identical to any other, living now or at any time in the past. Even 'identical' twins, though they look very similar, are not actually identical in every way. There are many ways in which humans vary.

Look at variation in your class. Try to find at least 20 ways in which individuals in your class vary from one another. Do not include things like clothes and jewellery, but just variations in their bodies.

All the people in your class are of similar age. Yet they will vary quite a lot in size. This is not just their height, but the size of different parts of their body, too.

It's not just humans, either. Every population of every species shows variation. All dandelions may look the same to you, but closer inspection would reveal a host of differences between individual plants.

> **Discussion point**
>
> Continuous variation is caused by the effects of multiple genes acting together. Why is it highly unlikely that a single gene could cause continuous variation?

▶ What types of variation are there?

Scientists describe variation as either **continuous** or **discontinuous**. Continuous variation is where there is a continuous range with no 'categories' (for example, height in humans; people can be any height within a certain range) whereas in discontinuous variation there are distinct groups (for example, fingerprint types as shown in Figure 10.1; a person's fingerprint can be one of an arch, a whorl or a loop – there are no 'in-between' fingerprints).

Figure 10.1 Finger print groups: a) an arch, b) a loop, c) a whorl. This is an example of discontinuous variation.

Investigation into variation in organisms

Variation in finger length

Procedure

Collect data on the length of the middle finger of everyone in your class. Measure the finger length as shown in Figure 10.2.

In each case, record whether the person is male or female – we can then ask the question 'Is there any difference between the length of fingers in boys and girls?'

The data can be plotted as a bar chart in several different ways, as shown in Figure 10.3.

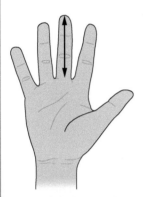

Figure 10.2 Measure the middle finger from the tip to the bottom of the crease where the finger joins the palm.

Analysing your results

1 Which would be the best way of plotting the data:
   a) if your main interest was in the data for the group as a whole, but you had some interest in differences between boys and girls
   b) if you were particularly interested in differences in the pattern between boys and girls?
   Give reasons for your answers in each case.
2 There are three possible hypotheses in connection to our question, 'Is there any difference between the length of fingers in boys and girls?'
   • There is no difference in finger length between boys and girls.
   • Boys tend to have longer fingers than girls.
   • Girls tend to have longer fingers than boys.
   a) From your data, which hypothesis is supported by the evidence?
   b) How conclusive do you think this evidence is? Give reasons for your answers.

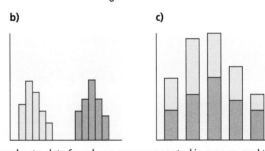

Figure 10.3 In each of these bar charts, data from boys are represented in orange, and those from girls are in yellow. a) Boy and girl data plotted alongside each other, b) Boy and girl data plotted separately (could also be on two graphs), c) Boy and girl data combined but distinguished.

▶ What causes variation?

Variation is caused by two factors. The genes of an organism control its characteristics and different sets of genes will result in heritable variation. In humans, the only people with identical genes are identical twins (or triplets, and so on.) because they are formed from the splitting of a single fertilised egg cell. Yet there are variations even between identical twins (Figure 10.4). This is environmental variation, and is caused by the influence of the environment – resulting sometimes from unplanned life events (such as scars from wounds), and sometimes from the individual's personal choices (hair styling, body piercings or tattoos, for example). Occasionally, one twin is born much smaller than the other, because for some reason it was deprived of nourishment in the mother's uterus (Figure 10.5) – this is another example of environmental variation.

Figure 10.4 Identical twins have identical genes but there are still some differences between them.

Figure 10.5 These two babies are identical twins but there was a complication in pregnancy that resulted in blood flowing from the smaller one to the larger one, so the smaller twin received less oxygen and nourishment. This is an example of environmental variation.

Some variations might result from a combination of genetic and environmental factors – height and weight, for example, have genetic components but are also affected by diet.

> ### ✔ Test yourself
>
> 1 What are the causes of variation (heritable, environmental or both) in the following features?
>   a) height
>   b) eye colour
>   c) skin colour
>   d) body piercings
> 2 Define the term 'continuous variation'.
> 3 Classify the following as demonstrating either continuous or discontinuous variation.
>   a) weight
>   b) length of foot
>   c) shoe size
>   d) hair colour

Why don't you look exactly like your parents?

Offspring are genetically different from their parents as a result of sexual reproduction, which involves an egg fusing with a sperm in the process of fertilisation. The genes from the mother in the egg are mixed with different genes from the father in the sperm. The cell formed as a result of fertilisation (the zygote) has one set of genes from the father and one set from the mother. The 'set' of genes in a gamete represents only half of the mother's or father's total number of genes, and the combination of genes making up the 'set' varies, which is why brothers and sisters are similar but different.

Organisms that reproduce by asexual reproduction do not mix their genes because fertilisation does not take place. One individual produces offspring that are genetically identical to each other and to the parent. These are called clones.

Key terms

Sexual reproduction Reproduction involving fertilisation (the fusion of female and male gametes), which results in offspring that are genetically different to the parent(s).

Fertilisation The joining of a male and female gamete.

Asexual reproduction Reproduction that does not involve fertilisation, and which results in offspring that are genetically identical to the parent.

Gamete A sex cell (for example, sperm, egg, pollen).

Figure 10.6 A set of identical (clone) plants, grown from cuttings taken from a single parent plant.

Key term

Mutation The creation of new genes.

The genes of a species do not remain the same for all time. New alleles and characteristics are constantly appearing. For example, it is likely that there were no blonde prehistoric humans. At some point in human history, a 'blonde allele' must have appeared. Changes to genes are caused by mutation. A mutation is a change in the structure of a gene. These changes occur naturally, randomly and often, but the rate of mutation is increased by ionising radiation and by certain chemicals in the environment. Most mutations make such small changes that no effect is seen. Some are harmful, but very rarely a mutation can arise that actually 'improves the design' of the organism and helps its survival.

Figure 10.7 It is thought that prehistoric humans were mainly dark haired and dark skinned (although new evidence indicates that some may have been redheads).

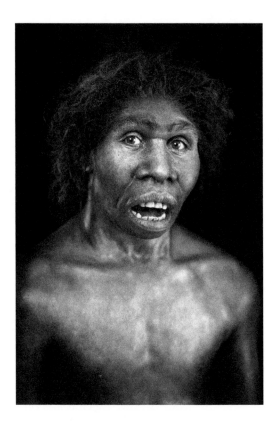

▶ What is an inherited disease?

Some mutations can result in an allele that is harmful and causes a disease. This allele, and therefore the disease, can be inherited. An example of this is **cystic fibrosis** – the lungs and digestive system of people with this disease become clogged with thick mucus, which makes breathing and digesting food difficult, and leads to reduced life expectancy.

The cystic fibrosis allele is recessive, so the disease only appears when an individual has the cystic fibrosis allele for this gene on both chromosomes. There are about 8500 cystic fibrosis sufferers in the UK but over 2 million people 'carry' the faulty allele. That means they have one cystic fibrosis allele and one 'normal' allele. Someone who is heterozygous for a recessive trait, such as cystic fibrosis, won't suffer from the disease but may pass it on to any children. This person is called a carrier. If two people carrying the cystic fibrosis allele have children, there is a 1 in 4 chance that any child they have will suffer

10 Variation and evolution

from the disease. This can be shown using a Punnett square, as in Chapter 9. Let us call the normal allele **C** and the cystic fibrosis allele **c**.

Parent genotype: mother × father
                   Cc   **Cc**
Parent phenotype: normal   normal
Gametes: **C** or **c**   **C** or **c**

		Male gametes	
		C	c
Female gametes	C	CC	Cc
	c	Cc	cc

A child with alleles **cc** will suffer from cystic fibrosis.

It is likely that around half of the couple's children will carry the disease but not suffer from it.

Inherited diseases and family trees

Examination of a family tree can indicate how a disease like cystic fibrosis is inherited. Some interpretation is needed when the disease is caused by a recessive allele, as the heterozygous carriers appear no different to the healthy individuals who are not carriers. A worked example is shown below.

Figure 10.8 shows a family tree of a family with cystic fibrosis sufferers. Interpreting the family tree, we can get some idea of the genotypes of most, but not all the individuals.

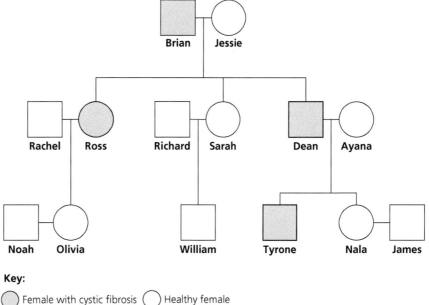

Key:
(circle, shaded) Female with cystic fibrosis   (circle) Healthy female
(square, shaded) Male with cystic fibrosis   (square) Healthy male

Figure 10.8 Family tree for family including cystic fibrosis sufferers.

Let's call the normal (dominant) allele **C** and the cystic fibrosis (recessive) allele **c**.

▶ **Individuals with cystic fibrosis (genotype cc)** – We know that **Brian, Ross, Dean** and **Tyrone** have genotype **cc**, because we are told they have cystic fibrosis. As the allele is recessive, anyone with the disease must have genotype **cc**.

- **Carriers of cystic fibrosis (genotype Cc)** – As the cystic fibrosis allele is recessive, and people with the disease are homozygous, they will always pass on a copy to their children. Therefore all children of a sufferer will either have the disease, or will be a carrier. This means that **Sarah, Olivia** and **Nala** are carriers and have the genotype **Cc**. To get the disease, though, both parents have to contribute a **c** allele. And so **Jessie** and **Ayana** must be carriers, as well.
- **Clear of cystic fibrosis (genotype CC)** – Here, we don't really know. Sarah and Richard have a child who is healthy, and Rachel and Ross also have a healthy child. That does not prove that **Richard** and **Rachel** are not carriers though, as there not enough children to be confident. **William** will either have the genotype **CC** or **Cc**. We know nothing about **Noah** and **James**.

What we can tell from this family tree is summarised in Figure 10.9.

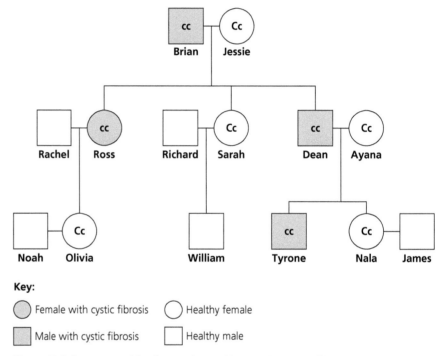

Key:

( ) Female with cystic fibrosis   ( ) Healthy female

[ ] Male with cystic fibrosis   [ ] Healthy male

Figure 10.9 Genotypes of family members, with regard to cystic fibrosis.

▶ What is gene therapy?

Gene therapy is the name given to a range of techniques that can be used to remove the effects of a harmful allele like the one causing cystic fibrosis. It can be done in several ways.

- **Introducing a 'healthy' allele into the person's DNA** – If the harmful allele is recessive, a healthy dominant allele will counteract it. The recessive allele does not have to be removed.
- **'Switching off' the harmful allele** – This can be done in various ways, including the introduction of a completely new gene into the body.

Gene therapy has proved very difficult to carry out and is still in the course of development.

It was reported in 2015 that, for the first time, researchers had been able to insert healthy alleles into the lungs of people with

Test yourself

4 Why does asexual reproduction not result in variation, but sexual reproduction does?

5 What is a mutation?

6 State two environmental factors that can increase the rate of mutation.

7 Why is it difficult to identify a carrier of a genetic disease caused by a recessive allele?

8 Define the term 'gene therapy'.

Key term

Evolution The process by which living species have gradually changed and developed from earlier forms over a long period of time.

cystic fibrosis. However, the success rate was low (those treated only showed a 3.7% improvement in lung function) and there is still a long way to go before any gene therapy treatment for cystic fibrosis will be generally available.

There are some issues around gene therapy. The research to develop treatments is very expensive and pharmaceutical companies may find it difficult to make any money. The only gene therapy drug available so far, to treat a rare genetic disease, has been priced at £810 000 per patient. Some religious groups believe that humans should never alter the genes of living organisms. On the other hand, the benefits could be enormous. For example, it offers the hope of allowing cystic fibrosis sufferers a normal length and quality of life.

Why is variation important for evolution?

Without mutations, living organisms would never change. The very first living cells would have been stuck with the genes they had, and could never have evolved into the billions of species that now inhabit the Earth.

We have already seen that organisms are well adapted to their particular environment, and that all populations show variation. These two ideas are closely connected, because without heritable variation, living things could never become adapted to their environment. How they do this is explained by the theory of evolution.

In a polar environment, nearly always snow-covered, it is best for animals to be white (Figure 10.10). They will be camouflaged and this will allow them to avoid being eaten or, if they are a predator,

Figure 10.10 Polar bears have evolved to be perfectly suited to a frozen environment.

to be able to approach their prey without being noticed. Because of variation, any animal population will contain individuals of various 'shades', some darker than others. In a polar environment the paler animals will survive better than the darker ones because they will be better camouflaged. More of them will survive to breed, and they will then pass on their 'pale' genes to their offspring. In this way, there will be more of the paler individuals in the next generation. This process will continue in each generation and the population will become paler and paler until all the individuals will be basically white (although there will always be some variation in colour).

Over long periods of time, animal and plant populations change in ways that make them better adapted to their environment. This gradual change is called evolution. If the environment changes significantly for some reason then the process may be re-set, and new adaptations may evolve for the new conditions.

▶ What is the theory of natural selection?

The theory of natural selection describes a mechanism by which evolution is thought to occur. It is one of the most famous theories in science, and was originated by Charles Darwin (see Figure 10.11).

In Darwin's time, many people believed that God had created every species separately and that one species never changed to give rise to new ones. Others believed in evolution but thought that changes came about by what the organism did, or what happened to it, in its lifetime. Darwin went on a five-year scientific voyage of discovery on the ship H.M.S. *Beagle* between 1831 and 1836. He discovered many new species and noticed that, often, different species were variations on what seemed to be a common basic model. What was more, the variations all seemed to be linked with the organism's environment or lifestyle (see Figure 10.12). Darwin could not do experiments on evolution, which can take thousands of years. All he could do was to make careful observations and then try to devise hypotheses to explain them.

Figure 10.11 Charles Darwin, the famous British naturalist.

Figure 10.12 Variations in Galapagos tortoises. The one on the left has a domed front to its shell and a longer neck. It lives on an island with little ground vegetation so it has to reach up to feed on bushes. Its shell and neck adaptions allow it to do this.

tree finches	fruit eaters	parrot-like bills	 *Camarhynchus pauper*		
	insect eaters	grasping bills	 *Camarhynchus psittacula*	 *Camarhynchus parvulus*	 *Camarhynchus pallidus*
ground finches	cactus eaters	probing bills	 *Geospiza scandens*		
	seed eaters	crushing bills	 *Geospiza difficills*	 *Geospiza fuliginosa*	 *Geospiza magnirostris*

Figure 10.13 Galapagos finches, showing variations in beak shape and size according to diet.

He noticed that there were species of finch that were only found on certain islands in the Galapagos archipelago. They were similar to each other, and to a type of finch found on the South American mainland about 500 miles away, but each one had its own specific characteristics. In particular, the birds' beak shape and size seemed to reflect the food they ate. This is shown in Figure 10.13.

At the time, many people believed that characteristics were acquired during an organism's lifetime and then passed on to their offspring. For example, it was thought that if you spent a lifetime in a physically demanding job, your children would be more muscular. There is no known mechanism that would cause a beak to get bigger and stronger through trying to crush seeds, as the beak is not made of muscle. There is also no way that probing into bark for insects would cause a beak to get thinner and therefore better at probing. Darwin's observations of the Galapagos finches did not fit this hypothesis.

For the next year or two, Darwin thought about what he had seen and eventually developed his theory of natural selection to explain the evidence. This was as follows:

▶ Most animals and plants have many more offspring than can possibly survive, therefore the offspring are in a sort of 'battle' for survival. This is the idea of **over-production**.
▶ The offspring are not all the same; they show **variation**.
▶ Some varieties must be better equipped for survival than others, because they are 'better fitted' to the environment. These will be more likely to survive to breed (that is, '**survival of the fittest**').
▶ Those that survive will **breed** and pass on their heritable characteristics to the next generation (Darwin did not know the details of this, because in his time people did not know about genes).
▶ Over many generations, the best characters will become more common and eventually spread to all individuals. The species will have changed, or **evolved.**

Darwin named his theory the theory of natural selection, but he was not the only scientist working on a theory for the mechanism of evolution at the time. Another was Alfred Russel Wallace (Figure 10.14) who, working independently of Darwin, came up with virtually identical ideas. The two men decided to cooperate and together produced the first ever publication on natural

Figure 10.14 Alfred Russel Wallace.

selection in 1858. Although less well known than Charles Darwin, Alfred Russel Wallace is now widely regarded as the co-discoverer of natural selection. The theory of natural selection has been slightly refined over time, but is still accepted as the mechanism for evolution by the vast majority of scientists.

 Practical

Modelling evolution

It is difficult to conduct experiments on evolution because the process often takes thousands of years. When scientists cannot carry out experiments directly on living organisms for some reason, they can sometimes 'model' the process they wish to investigate. We can do this to investigate the way in which camouflage evolves.

Apparatus

> 100 plain wooden cocktail sticks
> 100 cocktail sticks coloured green
> stopwatch

Procedure

1 Count out 20 plain and 20 green cocktail sticks.
2 Mark out an area of long grass 1 m × 1 m.
3 One person in the group should now scatter the cocktail sticks across the grass in the area. It is better if the sticks are well spread, not clustered together.
4 A different person in the group now picks up cocktail sticks for 15 seconds. If 20 sticks are picked up before the time is up, stop. The person is acting as a predator and the collected cocktail sticks have been 'eaten'.
5 Draw a table to record your results in.
6 The remaining sticks in the area now 'breed'. Work out how many green and plain cocktail sticks remain, and double the number of each, by scattering new cocktail sticks back in the marked area. For example, if there are 12 green sticks left and 8 plain, then add another 12 green sticks and 8 plain sticks into the area. Record the new numbers of green and plain cocktail sticks.
7 Repeat steps 4 and 6 four more times, or until the population would exceed the number of cocktail sticks you have available.
8 Plot your results as a bar chart.

Analysing your results

1 What happens to the green and plain cocktail stick 'populations' over the 'generations' in this experiment?
2 Why is it suggested that you plot the data as a bar chart rather than as a line graph?

 Discussion point

In this model you are modelling the behaviour of a predator and the cocktail sticks are the prey. How accurate do you think this model is? What are its limitations?

▶ **Extinction – a failure of natural selection?**

Millions of species that existed in the past are no longer found on Earth – they have become extinct. This could happen for a variety of reasons:

1 The organism has failed to adapt successfully to its environment.
2 The organism has adapted to its environment to some extent, but another similar organism has adapted better. The less successful organism cannot compete and eventually dies out.
3 The organism has adapted to its environment well, but the environment suddenly changes and the organism cannot survive in the new conditions.

The first reason (complete failure to adapt) is virtually unknown. It might explain the extinction of a new mutation, but not a whole species.

The second reason is more common. For example, since the 1990s numbers of white-beaked dolphins have been declining around the Scottish coast, and this has been linked with an increase in numbers of another species, the short-beaked dolphin. Warmer sea temperatures have encouraged the short-beaked dolphin to move into the area from further south, and it is thought that it might be 'out-competing' the white-beaked form (Figure 10.15).

Figure 10.15 a) White-beaked dolphin, b) short-beaked dolphin. The two species are competing in the seas around Scotland.

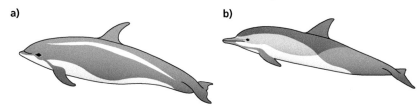

a)   b)

Sometimes, problems can arise from the introduction of a species that was not present before – an 'alien' species. For example, 150 years ago the red squirrel was common throughout Wales, but in the late 19th century the grey squirrel was introduced into the UK from the USA. It lives in similar places to the red squirrel but is better adapted and so the red squirrel has gradually disappeared from areas where the grey squirrel is present. Red squirrels are now absent from South and West Wales, and are restricted to a few areas in mid and North Wales (Figure 10.16). Red squirrels are not going to become extinct, because they thrive in other places around

Figure 10.16 The distribution of the red squirrel in Wales.

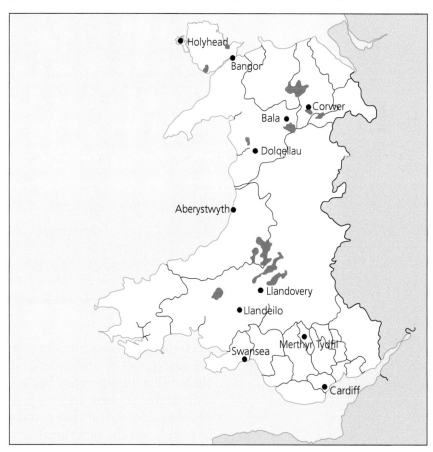

Figure 10.17 The dodo, which became extinct at the end of the seventeenth century – less than 100 years after its discovery.

Europe, but if they did not have such a wide range, they would be threatened with extinction.

The third reason is a common cause of extinction and is often linked with the activity of humans. A famously extinct animal is the dodo, a flightless bird that used to live on the island of Mauritius (see Figure 10.17). It had adapted successfully to its environment and had no natural predators. When Dutch settlers colonised the island in 1638, they brought with them cats, dogs, rats (from the ships) and pigs. The dodos were easy prey for humans as they could not fly and had never needed to evolve to be cautious. Humans ate quite a lot of them (although they didn't taste particularly good) and the cats, dogs and rats also fed on their eggs and young. Within a century the dodo was extinct, its environment having changed completely with the introduction of predators. The problem is the speed of change. Natural selection is a slow process, and if there is a rapid change in the environment, as there was in Mauritius, species can become extinct before they get a chance to fully adapt.

How has natural selection led to 'superbugs'?

Natural selection is a constant and ongoing process. It usually takes a very long time, but in certain circumstances it can happen quite quickly. One example is the evolution of 'superbugs' (Figure 10.18). These are bacteria that are resistant to the antibiotics normally used to treat infections.

Figure 10.18 A variety of 'superbugs' have appeared that are very difficult to treat with antibiotics.

Bacteria show variation like every living organism. Most of them are susceptible to antibiotics, but when you take these medicines there will always be a few bacteria that are naturally resistant and will survive. If there are only a few of them in you, they will not cause problems, but they could still be spread to others. If lots of people use the same antibiotics, after a while the susceptible

Figure 10.19 Evolution of resistance to antibiotics in bacteria. Note that in reality many more generations would be required before full resistance evolved.

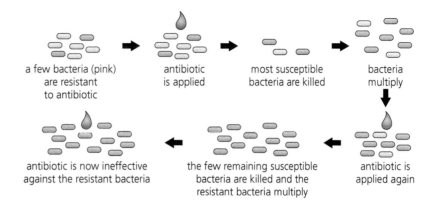

a few bacteria (pink) are resistant to antibiotic → antibiotic is applied → most susceptible bacteria are killed → bacteria multiply

antibiotic is now ineffective against the resistant bacteria ← the few remaining susceptible bacteria are killed and the resistant bacteria multiply ← antibiotic is applied again

bacteria are mostly killed off, leaving the resistant ones able to multiply. Given time, the whole population of bacteria will be resistant to the antibiotic (Figure 10.19). The bacteria have not become a new species but in a fairly short period of time they have evolved to be resistant to certain antibiotics by a process of natural selection. Natural selection is rapid in bacteria because they reproduce very rapidly (about once every 20 minutes, on average), and so in the course of a day they can progress through 72 generations! To avoid resistance building to new antibiotics, doctors try not to prescribe these drugs unless it is absolutely essential, and then try to use a variety of different antibiotics.

A similar problem has been encountered with some widely used pesticides, which have killed off most of the pests that are susceptible to it, leaving those that are naturally resistant to it to breed and spread.

Why is human genome mapping important?

The **genome** is the name given to all the genetic information in an organism. It includes all the genes and their sequence on all the chromosomes, and the DNA base pairs that make up those genes. The Human Genome Project was an international scientific research project that worked out the sequence of chemical base pairs in human DNA, identified all the genes (and their variants) in humans and their location on the chromosomes. It was completed in 2003. Mapping the human genome was a huge task and it has huge potential importance for medicine in the future. Some genes are known to directly cause disease – for example, there are mutations of certain genes that can cause cancer. Others can make a person more susceptible to a disease, while others play a key role in protecting the body. Knowing about the existence of these genes and their location on the human chromosomes allows the possibility of altering them or counteracting their effects. It also allows the possibility of creating targeted drugs or viruses that would (for example) only attack cells containing a mutated cancer-causing gene.

There appear to be about 20 500 genes in a human being (which is fewer than was previously thought), and many of them interact in complex ways. Mapping the genome is just the start and work is ongoing to discover exactly how these genes work, which could lead to further developments and applications.

✔ | **Test yourself**

9 What is the difference between evolution and natural selection?
10 Why would a population that showed no variation never evolve?
11 Which two scientists were responsible for introducing the theory of natural selection?
12 What is meant by 'survival of the fittest'?
13 Why is sudden environmental change much more likely to lead to extinctions than gradual environmental change?
14 What is a genome?

→ | Activity

Why has rat poison stopped working?

Warfarin was a common ingredient of rat poison but its use has declined because rat populations have become resistant to it, so its effectiveness has decreased. It is thought that this is because of natural selection. Rats that have a natural resistance to warfarin have survived and reproduced, whereas those that are susceptible have been mostly killed off.

Different rat populations have differing levels of resistance to warfarin. Scientists sampled rat populations from five sites (A–E) in a fairly wide geographical area and tested them for warfarin resistance. The results are shown in Figure 10.20.

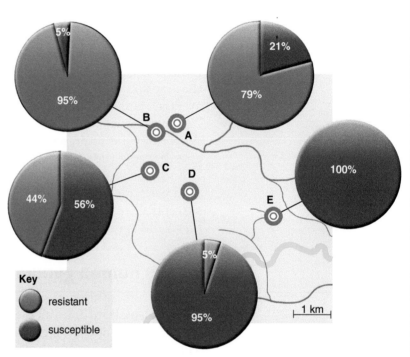

Figure 10.20 Results of warfarin-resistance tests on rat populations from five different sites (A–E).

Analysing the results

1 Look at Figure 10.20. Put sites A–E in order of resistance to warfarin (highest resistance to lowest resistance), as shown by the rat populations.
2 Do you think the differences in resistance in the five sites are significant? Justify your answer.
3 Look at the conclusions listed below. In each case, say whether you think the data support the conclusion and, if so, how strong the evidence is (with reasons).
   a) Warfarin has never been used at site E.
   b) The rat populations at the different sites are genetically different.
   c) Resistance to warfarin first arose in site B.
   d) Environmental conditions are different in sites D and E when compared with the other sites.
   e) Resistant rats have migrated from areas A and B to area C.

Chapter summary

- Variation in individuals of the same species can have environmental or genetic (heritable) causes.
- Variation can be classified as continuous or discontinuous.
- Sexual reproduction leads to offspring being genetically different from the parents, but in asexual reproduction the offspring are genetically identical clones.
- New alleles result from mutations, which are changes in existing genes; mutations occur at random and often have no effect, but some can be beneficial or harmful.
- Mutation rates can be increased by environmental factors such as ionising radiation.
- Some mutations cause inherited diseases such as cystic fibrosis.
- Gene therapy offers hope to cystic fibrosis sufferers and those with other heritable diseases.
- Heritable variation is the basis of evolution.

- Individuals with characteristics adapted to their environment are more likely to survive and breed successfully. Genes that have enabled these better adapted individuals to survive are then passed on to the next generation.
- The theory of natural selection was proposed by Alfred Russel Wallace and Charles Darwin.
- The process of natural selection is sometimes too slow for organisms to adapt to rapidly changing environmental conditions and so species may become extinct.
- Antibiotic resistance in bacteria, pesticide resistance and warfarin resistance in rats are examples of natural selection happening over a shorter than usual period of time.
- Understanding the human genome opens up many potential benefits in medicine, allowing new and targeted forms of treatment.

► Chapter review questions

1 The patterns of inheritance of cystic fibrosis in two families is shown as a family tree below.

Cystic fibrosis results from a homozygous pair of recessive alleles. People who are heterozygous for cystic fibrosis have one normal allele and one cystic fibrosis allele. They are carriers of cystic fibrosis but do not suffer from it.

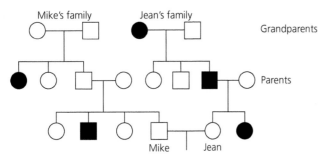

Key:

○ Female without cystic fibrosis
● Female with cystic fibrosis
□ Male without cystic fibrosis
■ Male with cystic fibrosis

a) In the family trees shown, if **N** = the normal allele and **n** = the allele for cystic fibrosis, what is the genotype of: [2]

  i) Mike's grandfather

  ii) Jean?

b) What is the percentage chance that Mike is a carrier of cystic fibrosis? [1]

Chromosomes from Mike and Jean's developing baby and from Mike were examined. A genetic analysis of the alleles present was carried out. The results are shown below as a sequence of bars.

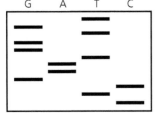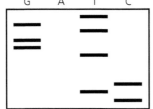

Genetic analysis of Mike's alleles   Genetic analysis of Mike and Jean's developing baby's alleles

c) What term is used for this sequence of bars? [1]

Cystic fibrosis is caused by a change in protein made in the cells.

d) Explain why the protein made in the cells of the developing baby is different from the protein being made in Mike's cells. [2]

e) Explain why genetic analysis is a more accurate method of predicting the inheritance of cystic fibrosis than using information from family trees. [2]

(from WJEC Paper B1(H), Summer 2014, question 8)

2 The European adder (*Vipera berus*) is a snake found in many parts of Wales. The body colour is usually brown, cream or red with a dark zigzag pattern along the back.

a) What word is used to describe the differences between members of the same species? [1]

b) Adders whose colour is all black are seen occasionally.

  i) What word is used to describe this sudden change in colour? [1]

  ii) What chemical is altered to cause this change in colour? [1]

c) State one way the offspring produced by sexual reproduction differ from those produced by asexual reproduction. [1]

(from WJEC Paper B1(H), January 2012, question 2)

3 In 1960, a population of wild rats in Wales was found to be resistant to the concentrations of warfarin that normally killed them. Scientists investigated the effect of warfarin on resistant and non-resistant rats.

They used samples of the population that showed resistance in the wild and equal numbers of laboratory-bred non-resistant rats. They noted the percentage that survived various concentrations of warfarin. The results are shown in the graph.

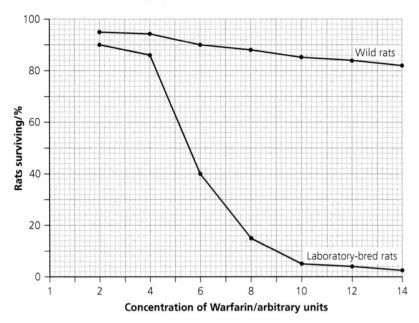

a) Compare the effect of increasing the concentration of warfarin on resistant and non-resistant rats. *[2]*

b) State the percentage deaths of:

i) resistant rats

ii) non-resistant rats

at a warfarin concentration of 10 arbitrary units. *[1]*

c) Explain how resistance to warfarin has evolved and spread in populations of rats since 1960. *[4]*

d) State one scientific reason in favour of using laboratory rats for this investigation. *[1]*

e) State one reason why animal rights supporters might not be in favour of using laboratory rats. *[1]*

(from WJEC Paper B2(H), Summer 2010, question 7)

11 Response and regulation

▶ If I stumble, will I fall?

We all trip up from time to time. Usually, we will stumble and right
ourselves, but occasionally we actually fall. Have you ever thought
about what is going on during this process?

When you trip, the first stage in avoiding a fall is to realise
that you've tripped. There are lots of signs – sense organs in your
ears detect that you are no longer vertical; your muscles and skin
may sense that you are not in contact with the ground; your eyes
may see the ground coming towards you! All this information is
sent to your brain, which then has to make a very rapid decision –
can you stop yourself falling? This decision involves all sorts
of factors like your speed and the angle of your body, and it's
important the brain gets it right, because the decision will affect
what it does next.

If you can correct the stumble, your brain must send out
signals that will re-balance your body. For example, if you are
falling forwards, it is best to put one leg forward to block any fall,
and lean your body slightly backwards to balance your forward
movement. If you are going to fall, putting your arms forward to
support you when you hit the ground will protect your face and
ribs (Figure 11.1), but this action would be counter-productive
if you were trying to stop yourself falling. All of these decisions
and actions are taken within a split second, and your brain nearly
always gets it right.

This is what your brain and nervous system do – they control
and coordinate the senses and responses in your body. And they do
it very well.

Figure 11.1 This footballer's brain has obviously decided he's going to hit the ground.

How does your brain get its information?

Information is fed to the brain constantly by a system of **sense organs** scattered around the body (Figure 11.2).

Sense organs are groups of special cells called **receptor cells**, which can detect changes around them, either internally or in the external environment. These changes are called **stimuli** (singular: stimulus) and include light, sound, chemicals, touch and temperature. Each group of receptor cells responds to a specific stimulus. The ears, for example, detect both sound and balance, but that is because they contain two different groups of receptor cells, one for sound and one for balance. The information from the sense organs travels to the brain and spinal cord (the **central nervous system**) along nerve cells (also called **neurones**).

Figure 11.2 Some of the body's sense organs.

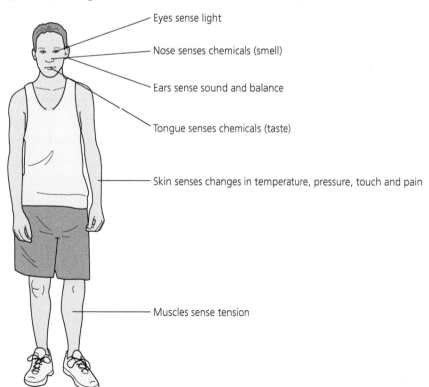

- Eyes sense light
- Nose senses chemicals (smell)
- Ears sense sound and balance
- Tongue senses chemicals (taste)
- Skin senses changes in temperature, pressure, touch and pain
- Muscles sense tension

How sensitive is your skin?

One of the things your skin can detect is touch. Scattered around in the skin are touch receptors, which detect if the skin is touched near them. If the skin is touched in between the 'detection areas' the body will not detect it (Figure 11.3).

Some areas of your skin have these touch receptors more closely packed than others, so that the sensitivity of your skin varies on different parts of your body. You can see how sensitive different areas of the skin are by an experiment. You will test how good the skin is at distinguishing two separate contacts that are very close together. Sensitive areas will detect that they have been touched twice. Less sensitive areas will only detect one touch.

Apparatus

> plastic-coated hairgrip
> ruler

Procedure

Work in pairs. One person will be the test subject, the other will be the experimenter.

1. Bend the hairgrip into a shape similar to that shown in Figure 11.4. By pressing on the sides, the gap between the points can be adjusted.
2. Set the points so that they are 10 mm apart. Ask the test subject to look away.
3. Touch the hairgrip to the skin on the test subject's fingertip 20 times. Sometimes, touch both points on the skin, sometimes just one point. Each time, the subject must say if they feel one point or two. Construct a suitable table to record your results. Record how many times the subject was right and how many times they were wrong.

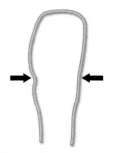

Figure 11.3 Diagram of the skin's surface showing areas covered by touch receptors. Any contact in the pink areas will be detected. Any contact in the spaces between the pink areas will not be detected.

Figure 11.4 A hairgrip unfolded for use in the experiment. Pressing as shown by the arrows can adjust the difference between the points

4. Repeat the test on two further areas of skin – on the palm of the hand and on the back of the hand.
5. Test all three areas again with the hairgrip points at different distances apart – 8 mm, 6 mm and 4 mm.
6. If time allows, the test subject and experimenter can swap roles and repeat the experiment.
7. Draw a suitable graph to display the data.

Analysing your data

8. What is your conclusion from the data? How strong is the evidence for this conclusion?
9. What are the possible sources of error in the experiment? Do the data indicate that any of these errors might have been significant? Is there anything that you could do to reduce these errors?
10. Why did you choose the type of graph you did? Would it have been possible to use any other sort of graph?

How does information travel in the body?

Your brain and spinal cord make up the **central nervous system (CNS)**. Together they coordinate and control your body – to do this, they need to receive information from the sense organs, and send out information to muscles in order to make a response happen. The information travels as electrical impulses along nerves. Together, the central nervous system and the nerves form the **nervous system**.

Some nerves take signals to the CNS from sense organs and other nerves take information out from the CNS to the body. So, when your body senses a stimulus, the sense organ concerned sends an impulse along a nerve to the CNS. Sometimes the message goes to the brain, which then decides what to do about it. If action is needed, the brain sends an impulse along a nerve to the appropriate part of the body, which then reacts. This reaction is called a **response**. The time taken between the stimulus and the response is known as the **reaction time**.

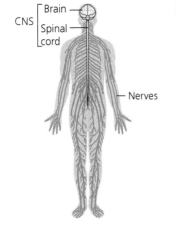

Figure 11.5 The human nervous system.

Specified practical

Investigation into factors affecting reaction time

This experiment looks at whether practice can improve reaction time. You can get a measure of reaction time by seeing how quickly someone catches a ruler that is dropped between their fingers. Our hypothesis is that **practice will reduce the reaction time**. Let's look at the evidence for this hypothesis, because a hypothesis is more than just a guess – it has to be built on scientific principles.

> It is known that the time taken for an impulse to travel along a nerve is fixed. Every time you catch the ruler, the signals will travel along the same nerves, so the overall time taken will always be the same. This is evidence against our hypothesis.
> Catching the ruler does not just involve a simple pathway. You may be able to anticipate that the person is going to drop the ruler from tiny 'signals'; you may be able to develop greater powers of concentration. These are skills that could be improved by practice. This is evidence for our hypothesis.
> We know that similar things can be improved by practice. For example, in cricket, fielders practise to improve their catching ability. Catching a cricket ball is a more complex action, yet this still provides evidence for our hypothesis.

The hypothesis suggested is a valid one. There is evidence to back it up, yet we don't know for certain that it is true. (If we did, there would be no point doing an experiment!)

Apparatus

> 30 cm ruler

Procedure

This is the basic method for working out reaction times. Work in pairs. One person is the experimenter, the other is the subject.

1 Mark a pencil line down the centre of the subject's thumbnail on the right hand if they are right-handed, or on the left hand if they are left-handed.
2 Ask the subject to sit sideways at a bench or table with their forearm resting flat on the bench and their hand over the edge.
3 Hold a ruler vertically between the subject's first finger and thumb with the zero opposite the line on the thumb but not quite touching the thumb or fingers (Figure 11.6). The distance between thumb and finger should be exactly the same for every trial.

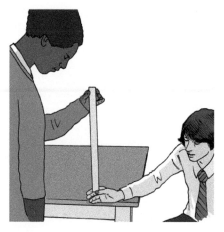

Figure 11.6 Carrying out the procedure.

4 Ask the subject to watch the zero mark and, as soon as you release the ruler, the subject must try to catch it between their finger and their thumb to stop it falling any further. Construct a suitable table to record your results. Record the distance on the ruler opposite the mark on the thumb.
5 Repeat this four more times (to give five in total) and calculate the average distance. Convert this distance to a time using the graph in Figure 11.7.
6 Now use this method to design and carry out an experiment to test the hypothesis that practice improves the reaction time in this exercise.

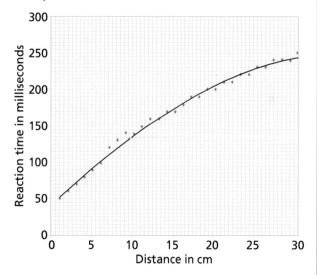

Figure 11.7 Conversion graph for converting catch distance into reaction time. This is necessary because the ruler accelerates as it falls.

Analysing your results

1 Write a full report of your experiment, describing your method, results and conclusions.
2 Evaluate your design. Could it be improved in any way?

How do we see?

Vision is a complex sense. For an animal's brain to build a picture of what is going on in front of it, it needs a lot of information. The sense organs that provide this information are the eyes. These sense organs must:

- be able to detect light and distinguish different intensities of light, in order to collect data on the patterns of light and dark
- be able to detect different colours, or at least different tones (not all animals can see colours)
- have a way of protecting the light-detecting surface, since vision is a vital sense to those animals that have it
- have some sort of focusing system so that objects at different distances can be seen clearly
- be able to adjust the level of light hitting the sensory surface, in order to work in different levels of brightness
- prevent light from being reflected on its way to the sensory surface, or else the image will be confused.

The eyes of mammals have features that cope with all these problems:

- The retina has light-sensitive cells that detect light, and some of them detect coloured light. As there are light-sensitive cells spread all across the retina, the pattern of light can be detected. The optic nerve at the back of the eye conveys nerve impulses from these cells to the brain, which analyses them to build up a picture of what is being seen. As there is no retina at the point at which the optic nerve leaves the eye, this part is referred to as the blind spot. You do not notice a blind spot in your vision, as the brain judges what the tiny spot would contain and 'fills it in'.
- The retina is protected by a tough outer coat, the sclera. Around most of the eye, this is white and opaque, but it has to be transparent at the front to let light in. The transparent part is called the cornea.
- The light coming into the eye is focused by the lens. The lens is flexible and can change shape to be able to focus objects at different distances.
- The amount of light coming into the eye through the pupil is adjusted by the coloured part of the eye, the iris. Muscles in the iris can make the pupil larger or smaller and so let more light in when it is dark, and less light when it is bright.
- Underneath the (transparent) retina is a black layer called the choroid. Black objects do not reflect light, and so this layer prevents light reflecting around the inside of the eye, and so prevents the light from being detected multiple times.

The structure of the human eye is shown in Figure 11.8, with an outline of the functions of the different parts.

Figure 11.8 Structure of the human eye.

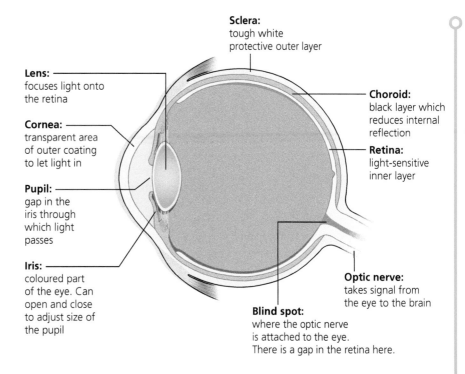

Lens:
focuses light onto
the retina

Cornea:
transparent area
of outer coating
to let light in

Pupil:
gap in the
iris through
which light
passes

Iris:
coloured part
of the eye. Can
open and close
to adjust size of
the pupil

Sclera:
tough white
protective outer layer

Choroid:
black layer which
reduces internal
reflection

Retina:
light-sensitive
inner layer

Optic nerve:
takes signal from
the eye to the brain

Blind spot:
where the optic nerve
is attached to the eye.
There is a gap in the retina here.

✓ Test yourself

1 For humans, a change in the temperature of a room would be regarded as a stimulus, whereas a change in the carbon dioxide levels would not. Explain.
2 What structures make up the central nervous system?
3 What stimulus does the retina detect?
4 Which part of the eye is responsible for focusing the image?
5 Which part of the eye reduces the chances of us getting dazzled by bright light?

▶ What is a reflex?

A simple example of how the nervous system coordinates the body can be seen in reflexes. A reflex is a particular type of response to a stimulus. To be called a reflex, a response must be **rapid** and **automatic**. In general, reflexes are protective in some way. Examples of reflexes include the following:

▶ knee jerk
▶ breathing
▶ sneezing
▶ coughing
▶ blinking
▶ swallowing
▶ pupil reflex (your pupil getting bigger in the dark and smaller in the light)
▶ withdrawal reflex (pulling away from a painful stimulus).

Some of these actions (for example, blinking) can be done on purpose. If you deliberately blink, it is not a reflex, but most of the time you blink without thinking about it, and in those circumstances it is a reflex action. Blinking is also unusual as it can be caused by two different stimuli, either irritation of the cornea (by drying out or by dust) or seeing something approach the eye quickly.

A reflex involves a stimulus, a receptor, a coordinator, an effector and a response.

- ▶ A stimulus is a change in the environment that can be detected.
- ▶ A receptor is an organ that detects the stimulus.
- ▶ A coordinator detects the signal from a receptor and sends a signal to the effector.
- ▶ The effector is the part of the body (usually a muscle) that produces the response.
- ▶ The response is the action carried out.

The way the withdrawal reflex works is shown in Figure 11.9.

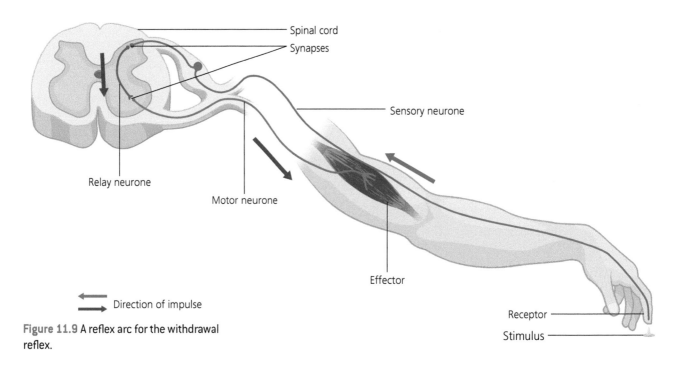

Spinal cord
Synapses
Sensory neurone
Relay neurone
Motor neurone
Effector
Receptor
Stimulus

Direction of impulse

Figure 11.9 A reflex arc for the withdrawal reflex.

Key terms

Neurone A nerve cell.
Sensory neurone A neurone that carries information from a receptor to the central nervous system.
Motor neurone A neurone that carries information from the central nervous system to an effector.
Relay neurone A neurone that transmits an impulse from a sensory to a motor neurone.
Effector A structure (a muscle or a gland) that carries out an action when stimulated by a motor neurone.

The stages in a reflex are as follows:

1 The stimulus is received by the receptor.
2 An impulse is sent along the sensory neurone to the spinal cord.
3 The impulse moves across a tiny gap (synapse) to the relay neurone.
4 The relay neurone transmits the signal (via another synapse) to the motor neurone.
5 The motor neurone stimulates the effector (the muscle) to respond.

In this reflex the coordinator is the spinal cord, as it is within the spinal cord that the nerve signal from the receptor is detected and transferred to the motor neurone, leading to the effector.

The impulse is automatic because it does not go through the brain, which is the part of the nervous system that 'makes decisions'.

The impulse is quick because it is the synapses that slow down an impulse, and in a reflex the impulse only has to go through two. If it went through the brain, it would go through thousands of synapses.

▶ Do plants respond to stimuli?

Plants do not move around, so you might think that they do not respond to changes in the environment, but they do. Their responses are slow and often involve growth towards, or away from, a stimulus. These growth 'movements' are called tropisms. There are several different kinds. Two examples are:

- ▶ Phototropism – This is growth in response to light. Plant shoots grow towards the light (positive phototropism) and roots grow away from light (negative phototropism).
- ▶ Gravitropism – This is growth towards the pull of gravity. Roots of plants show positive gravitropism and stems show negative gravitropism.

These responses ensure that plants always grow the right way up, whatever direction their seed lies in the soil. They also make sure that stems grow towards the light they need for photosynthesis (Figure 11.10), and roots always grow in a direction that enables them to get water, below the surface of the soil.

Plants don't have nerves, so these responses are brought about by special chemicals called hormones. Hormones are produced in response to a stimulus and travel to another part of the plant, where they cause a response. Animals also have hormones, as we shall see in the next section.

The specific hormone that causes tropisms in plants is called auxin.

Figure 11.10 These seedlings have grown towards the light from a nearby window.

> ### ✔ Test yourself
>
> 6 Which two characteristics of an action class it as a reflex?
> 7 What is the difference between a sensory neurone and a motor neurone?
> 8 In the blinking reflex, what is the effector?
> 9 Why is positive phototropism useful in plant stems?
> 10 What is the name of the plant hormone that causes tropisms?

▶ How do we keep conditions in the body stable?

For many animals, including humans, keeping certain conditions inside the body relatively constant is very important for survival, as it helps to ensure optimum conditions for the chemical reactions that are needed for life and the enzymes that control them. This regulation is called homeostasis. Hormones play a vital role in homeostasis.

Hormones are chemical messengers that are made in certain organs and travel around in the bloodstream, affecting various specific parts of the body. They are mainly used for medium-term and long-term regulation, whereas nerves generally control quicker responses.

The main conditions that are controlled by hormones in the body are described below.

Temperature

Temperature affects all chemical reactions and animals have a variety of ways of keeping their internal body temperatures relatively constant. More details are given later in this chapter.

Water content

The concentration of chemicals in the cells can affect the essential chemical reactions going on inside the body. All of these reactions take place in water, which is therefore essential for life. Too little water (dehydration) may make the body fluids too concentrated and damage the body, but too much water can also be dangerous, as it dilutes the body fluids. The concentration of our bodily fluids is maintained within safe limits by hormones.

Glucose levels

Glucose is a very important chemical in the body. It is the main source of the body's energy, but it can damage cells if present in high concentrations, so its level must be kept within a safe range. If blood glucose levels get too high following a meal, they can be reduced by the hormone insulin (a protein), which is released by the pancreas. Insulin is released into the bloodstream, where it converts soluble glucose to an insoluble carbohydrate called glycogen, which is stored in the liver.

Some people have a condition that means their body produces little or no insulin. If untreated, their blood glucose levels become dangerously high. This condition is called diabetes. Activities also influence blood glucose levels. Eating carbohydrates raises blood glucose, while exercise lowers it.

▶ What happens when glucose control goes wrong?

In type 1 diabetes (the most common type in young people), the body stops producing insulin. It is thought that type 1 diabetes may be brought on by the body over-reacting to a certain type of virus, with the result that the immune system destroys its own insulin-producing cells in the pancreas. As a result, blood glucose levels go up and up and the body tries to get rid of the excess glucose in the urine. The following symptoms are noticed:

- ▸ The blood glucose levels rise higher than would ever be seen in a healthy person.
- ▸ Glucose appears in the urine (a doctor has to test for this).
- ▸ The person passes a lot of urine because the body is trying to 'flush out' the glucose.
- ▸ Because of all the water that is also being lost in the urine, the person gets very thirsty.
- ▸ The body cannot actually use the glucose in the blood without insulin, so the patient feels very tired.

Doctors diagnose diabetes by the presence of glucose in the urine, because all of the other symptoms can be caused by other things apart from diabetes.

If diabetes is not treated, the blood sugar level will become so high that the person will die. It cannot be cured but it can be managed so that the sufferer remains otherwise healthy. The treatment of type 1 diabetes consists of three things:

- ▸ The person has to inject themselves with insulin (usually before every meal) to replace the natural insulin that is no longer being produced (Figure 11.11).

Figure 11.11 People with type 1 diabetes have to inject themselves with insulin several times a day.

- The diet has to be carefully managed. The patient has to eat the right amount of carbohydrate (which is the source of glucose) to match the amount of insulin injected.
- The patient usually tests his or her blood glucose levels several times a day, to make sure the level has not gone too high or too low.

There is another type of diabetes (type 2), which is more common in older people. It is not caused by a lack of insulin, but the body no longer responds properly to the insulin produced. It is milder and can usually be controlled by drugs such as metformin tablets, or even by just being careful with the diet. Type 2 diabetes tends to be associated with being overweight or obese and the incidence of type 2 diabetes is increasing quite dramatically in the UK at present.

What is negative feedback?

Control of blood glucose levels in healthy people is an example of a negative feedback mechanism. A rise in blood glucose sets in motion a series of events that results in the level being lowered again. In the same way, low blood glucose causes a process that raises the level. This mechanism is an example of negative feedback, and is summarised in Figure 11.12. It involves insulin and another hormone from the pancreas, glucagon, which raises blood glucose.

Negative feedback is an effective method of maintaining a factor at a relatively constant level, and several other conditions in the body are controlled by this sort of mechanism.

> **Key term**
>
> Negative feedback A mechanism whereby a change in a factor sets off a series of events that lead to that factor being brought back to the normal level.

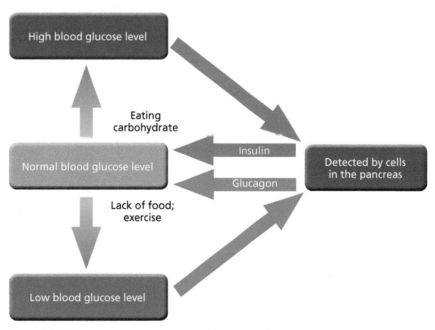

Figure 11.12 Negative feedback control of glucose levels.

► How is body temperature controlled?

Animals are kept alive by a series of chemical reactions that take place in their cells. These reactions are controlled by enzymes, which are affected by temperature. If the body temperature is not kept constant, essential reactions could stop. Mammals and birds control their body temperature precisely using various mechanisms. In humans, the body temperature is maintained at approximately 37 °C, but in other animals the normal body temperature can be different. Animals other than mammals and birds have to use different means to keep their body temperature within narrow limits – for example, moving into sunlight or shade to warm up or cool down. These mechanisms are not so precise and so their body temperature varies more than it does in a human.

The skin and temperature control

The skin is a complex structure that contains several different types of sense organ (Figure 11.13). It also performs various actions that help to control temperature.

Figure 11.13 A section through the human skin.

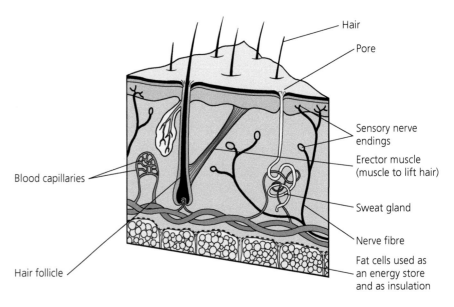

- Hair
- Pore
- Sensory nerve endings
- Erector muscle (muscle to lift hair)
- Sweat gland
- Nerve fibre
- Fat cells used as an energy store and as insulation
- Blood capillaries
- Hair follicle

► It produces sweat when the external temperature is high. The heat of the skin is used to evaporate the sweat and this in turn cools the skin.

► In hot weather, blood vessels near the surface of the skin dilate (get wider). This causes more blood to flow through them, so heat is lost to the atmosphere and the body cools down. In cold weather the vessels constrict (get narrower), so less blood goes to the surface to stop you getting colder. This is why you tend to go red when you're hot and pale when you're cold.

► The hairs in the skin stand on end in the cold to provide a thicker insulating layer to keep heat in. This isn't particularly effective in humans, because we don't have that much hair, but it's important in other mammals.

► When it gets cold, you shiver. The muscle contractions produce heat and so warm the blood a little.

These actions are illustrated in Figure 11.14.

The skin in cold conditions

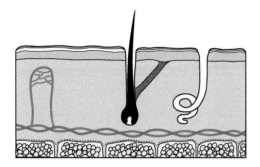

- Surface blood vessels contract. Blood takes a path in the skin far from the surface.
- Sweat glands stop producing sweat.
- Hairs pulled up by erector muscles contracting so layer of air trapped against skin surface is thicker. This provides insulation rather like double glazing.
- Shivering occurs by rhythmic contraction of skin muscles, which makes heat as a by-product.

The skin in hot conditions

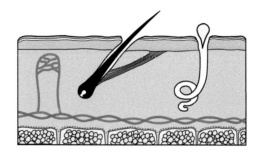

- Surface blood vessels relax. Blood takes a path in the skin near to the surface.
- Sweat glands produce more sweat, which cools the skin down as it evaporates from the surface. (We can produce one litre of sweat per hour.)
- Hairs drop down against skin surface as erector muscles relax so less air is trapped for insulation.
- No shivering.

Figure 11.14 Temperature regulation.

The control of body temperature is another example of negative feedback. A change in temperature sets in motion a series of actions that result in the temperature returning to the normal level again.

How can lifestyle affect regulation?

We have already seen that lifestyle choices can affect the risk of developing type 2 diabetes. The human body is incredibly complex and anything that disrupts any of the chemical reactions that make up life can have health effects. In type 2 diabetes the mechanism is not yet clear, but it seems that being overweight somehow alters the body chemistry and results in cells being less able to respond to insulin.

Drugs have multiple effects on the chemical reactions in the body. A drug is defined as a substance that alters the way the body works. While some of these changes can be beneficial over the short term, long-term use or abuse can create damaging side effects.

Remember that alcohol and the nicotine in tobacco are drugs, and their effects on body chemistry is what makes them harmful to health, too.

> **✔ Test yourself**
>
> 11 How do animal hormones travel around the body?
> 12 Where in the human body is insulin produced?
> 13 In which organ is carbohydrate stored, in humans?
> 14 A type 1 diabetic is going for a run one morning. At breakfast time, he takes slightly less than his usual dose of insulin. Why?
> 15 What is the difference between type 1 and type 2 diabetes?
> 16 How does sweating help to maintain our body temperature?

Chapter summary

- Sense organs are groups of receptor cells that respond to specific stimuli.
- The sense organs relay information as electrical impulses along neurones to the central nervous system.
- The brain, spinal cord and nerves form the nervous system; the central nervous system consists of the brain and spinal cord.
- The eye contains the following structures: sclera, cornea, pupil, iris, lens, choroid, retina, blind spot and optic nerve. Each of these plays a specific part in vision.
- Reflex actions are fast and automatic and some are protective.
- A reflex arc always has the following components: stimulus, receptor, coordinator and effector.
- A reflex arc consists of: a receptor, a sensory neurone, a relay neurone in the spinal cord, a motor neurone, an effector and synapses.
- Plant shoots respond to light (phototropism) and plant roots to gravity (gravitropism).
- Phototropism is due to a plant hormone, called auxin.
- Animals need to regulate the conditions inside their bodies to keep them relatively constant.
- Hormones are chemical messengers, carried by the blood, which control many body functions.

- In the human body, glucose levels need to be kept within a constant range.
- When blood glucose levels rise, the pancreas releases the hormone insulin (a protein) into the blood, which causes the liver to reduce the glucose level by converting glucose to insoluble glycogen and then storing it.
- Diabetes is a common disease in which a person is not able to adequately control blood glucose level.
- Type 1 diabetes is caused by the body not producing insulin. It is treated by injections of insulin.
- Type 2 diabetes is caused by the body cells not properly responding to the insulin that is produced. It is treated by controlling the diet and by tablets.
- The following processes contribute to the control of body temperature: change in diameter of blood vessels near the skin, sweating, erection of hairs, shivering.
- Negative feedback mechanisms maintain optimum conditions inside the body.
- Some conditions are affected by lifestyle choices, such as poor diet, over-consumption of alcohol and drug abuse. These affect the chemical processes in people's bodies.

▶ Chapter review questions

1 When Gareth heard a sudden loud noise, he jumped up out of his seat. His response was a reflex action.

   a) State two features that show it was a reflex action. [2]

   b) Copy the sentences, and fill in the blanks by using some of the words below. [3]

   eyes   sound   ears   light   receptors

   i) The stimulus in Gareth's reflex was _____ .

   ii) The stimulus was detected by _____ in Gareth's _____ .

   c) In what form is the information transferred to the brain? [1]

   (from WJEC Paper B1(F), January 2011, question 4)

2 A scientist carried out an investigation into the body temperature of a man. The changes in the man's body temperature were measured by a clinical thermometer in his mouth. The graph below shows his body temperature over a 35 minute period. Between 7 and 10 minutes he immersed his legs, from the knees downwards, in a bath of warm water at 40 °C. He then stepped out of the bath and dried his legs.

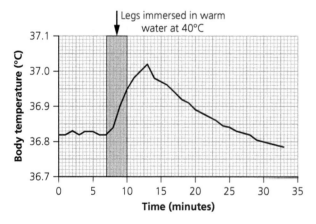

   a) Explain why the body temperature increased even though it was only the legs that were immersed in the warm water. [1]

   b) The experiment was repeated. After 20 minutes an electric fan was directed onto the man's legs. The results are shown in the graph below.

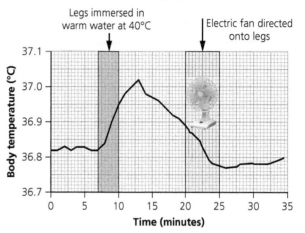

   Explain why the body temperature of the man dropped more quickly between 20 and 25 minutes when the fan was used. [2]

   (from WJEC Paper B1(H) new, January 2013, question 4)

3 In 2011 a contact lens was invented with a sensor that can measure the concentration of glucose in tears. It can be used to detect diabetes.

   a) Name two body fluids, other than tears, which can be tested to detect diabetes. [2]

   b) State three methods that are used to treat type 1 diabetes. [3]

   c) Which of the methods mentioned in part (b) is not normally part of the treatment for type 2 diabetes? [1]

   d) A poor diet in pregnant women increases the risk of their children developing diabetes. These children show abnormal development of cells in the pancreas. State two reasons why this could prevent the control of glucose concentration. [2]

   (from WJEC Paper B1(H) new, January 2013, question 7)

12 The kidney and homeostasis

 Specification coverage

This chapter covers the GCSE Biology specification section 2.6 Role of kidney in homeostasis.

It covers the structure and function of the kidney and its role in the regulation of the water content of the blood. The detail of the nephron is required along with the role of ADH. The treatment for kidney failure is also considered.

▶ What is the function of the kidneys?

We saw in the Chapter 11 that the body controls various factors like blood glucose and temperature within a restricted range, which allows the optimum working of living processes. The general term for such control is homeostasis. Another major factor that needs to be controlled is the water content of the body. All the chemical reactions that make up life take place in an aqueous solution. Differences in water content in body tissues affect concentrations and therefore rates of reaction, as well as determining the direction of water movement into and out of cells by osmosis. If the water content of the body gets too high or low it can have serious consequences. Dehydration, for instance, can be fatal.

The water content of the blood, and therefore of the body's cells, is regulated by the kidneys. These organs are responsible for the removal of waste products from the body. The wastes the kidneys remove are water soluble, and so their removal involves the loss of water, in urine. The kidneys control the water loss so that the concentration of the blood remains more or less constant, provided enough water is taken in via food and drink to replace the small amount of water that must always be lost in urine.

Urine contains the waste product urea, which is made in the liver by the breakdown of proteins that are not needed by the body. Urea is a poisonous substance and so cannot be allowed to build up and is removed from the blood by the kidneys.

The excretory system

The human excretory system is shown in Figure 12.1. Urine is formed in the kidneys and then drains down the ureters to be temporarily stored in the bladder, before exiting via the urethra. The blood is brought to the kidneys from the heart via the aorta and then the renal arteries, and then returns via the renal veins and the vena cava.

Key terms

Homeostasis The maintenance of a constant internal environment regardless of changing environmental conditions.

Aqueous solution A solution that has water as the solvent.

Dehydration A reduction in the normal water content of the body.

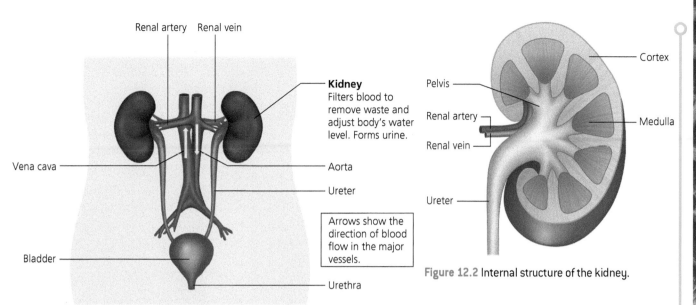

Figure 12.1 The human excretory system.

Figure 12.2 Internal structure of the kidney.

The internal structure of a kidney is shown in Figure 12.2. There are three main areas, the outer **cortex**, the inner **medulla** and (in the centre) the **pelvis**, which is where the urine drains before leaving via the ureter.

🧪 Practical

Dissecting a kidney

Apparatus

> kidney
> scalpel/scissors
> 2 mounted needles
> Petri dish

Procedure

1 Cut the kidney in half from top to bottom, using the scalpel or scissors.
2 Draw one half of the kidney to accurately show the location and proportions of the medulla, cortex and pelvis (Figure 12.3).
3 Cut a small piece of tissue (about 1 cm square) from the cortex.
4 Using the mounted needles, pull the tissue apart. Between the pieces, you should be able to see tubes of various diameters, some very thin. These are the nephrons that make up the kidney.
5 If available, use a stereo microscope to look at the kidney tubes.

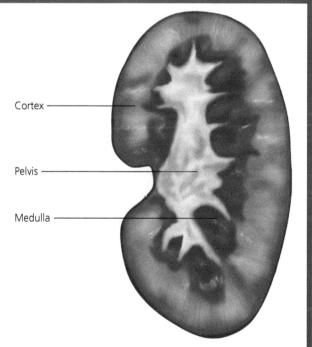

Figure 12.3 The appearance of a half kidney during dissection.

You are unlikely to have the ureter, renal artery and renal vein attached if the kidney has been bought from a butcher.

✓ │ **Test yourself**

1 What is the main waste product in urine?
2 Which organ manufactures this waste material?
3 Which blood vessel supplies the kidney with blood?
4 What tube connects the bladder and the kidney?

▶ How do the kidneys work?

A kidney is made up of millions of small tubes called **nephrons**, which extract wastes from the blood to produce urine, a waste solution containing urea and excess salts. The structure of a nephron is shown in Figure 12.4.

The walls of the **capillary knot** and the **Bowman's capsule** are leaky, and as blood flows through the capillary knot the blood pressure forces fluid through into the Bowman's capsule. The arteriole leading into the capillary knot is wider than the one leading away from it, so pressure builds up in the capillary knot. Only small molecules can get through the walls, so the walls act like a filter. Blood cells and larger molecules like proteins remain in the blood. (The presence of blood in the urine can indicate kidney disease, although there are other causes, as well.)

The materials that go through the filter into the nephron are water, glucose, urea and salts. Some of these substances are useful, however, and the body does not want to lose them. Urea is a waste product so it is a good thing to let that pass out in the urine, but glucose is a useful substance, and that is reabsorbed from the tubule into the blood. Some salts and most of the water are also reabsorbed into the blood, but how much is reabsorbed varies according to the body's needs.

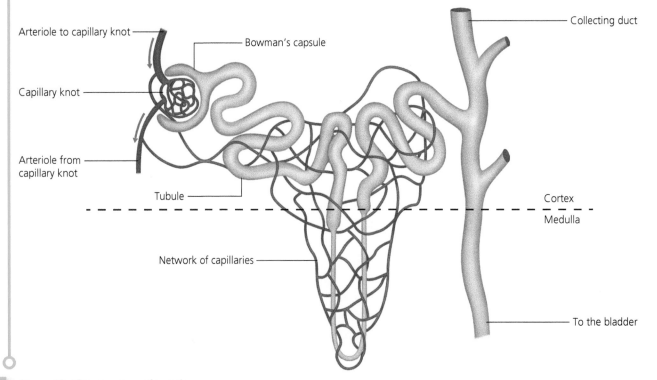

Figure 12.4 The structure of a nephron.

The process by which some things are reabsorbed and others are not is called selective reabsorption. Due to this reabsorption, the composition of the filtrate varies as it travels along the nephron tubule. By the time it reaches the end, the filtrate has changed into the liquid we call urine.

The amount of water reabsorbed in the nephron varies according to whether water is plentiful in the body, or in short supply. If the blood is too dilute, which can happen if the person has drunk a lot of liquid, less water is reabsorbed and the urine is pale and dilute. If the blood is too concentrated, then reabsorption is increased and the urine is darker and more concentrated. Normally, urine is most concentrated first thing in the morning, as you do not drink during the night while you are asleep.

The amount of water reabsorbed is controlled by a hormone, called anti-diuretic hormone (ADH). Areas in the brain detect the concentration of the blood and, if the concentration is too high, ADH is produced by the pituitary gland (a hormone-producing gland just beneath, but attached to, the brain). ADH causes the kidney to reabsorb more water and produce a more concentrated urine. Once the blood concentration returns to normal, ADH production stops and less water is reabsorbed.

Specified practical

Testing artificial urine samples for the presence of protein and sugar

Urine should not contain either sugar or protein. The presence of sugar is an indication that the patient has diabetes, and the presence of protein can indicate kidney disease (although in pregnancy a little protein is present in the urine – this is the basis for pregnancy tests).

You will be provided with artificial urine samples to test for protein and glucose, which is the sugar that would be present in diabetic urine.

The Biuret test for proteins and the Benedict's test for reducing sugars (glucose is a reducing sugar) are described in Chapter 3. Carry out each test on the artificial urine samples you are given. Note that the colours observed will be affected by the colour of the artificial urine sample. This will have little effect on the Benedict's test but may make it slightly more difficult to see the violet colour that indicates the presence of protein in the Biuret test.

Safety notes

Take care with hot water for the Benedict's test; use freshly boiled water from a kettle rather than boiling water over a Bunsen flame.

Wear eye protection, and wash off splashes onto the skin immediately.

▶ How can we treat kidney failure?

Most people have two kidneys, and if something happens to one – for example, if it has to be removed as a result of an accident or cancer – we can live a full and healthy life with only one working kidney. Indeed, some people are born with just one kidney, and live a normal life. If a patient has kidney disease, however, it is likely to lead to failure of both kidneys, and this is a life-threatening condition. There are two types of treatment for kidney failure:

- ▶ **Kidney dialysis** – The patient has to spend regular sessions attached to an artificial kidney machine, which removes wastes and restores the balance of salts and water in the blood.

▸ **Kidney transplant** – A new kidney can be placed in the body, so that kidney function is restored.

Each of these treatments has advantages and disadvantages, which we will look at below.

Kidney dialysis

When a patient is hooked up to a kidney dialysis machine, the blood is taken out of a blood vessel in the arm and pumped through the machine (Figure 12.5). A special dialysis fluid containing salts, called the **dialysate**, is also put through the machine. The dialysate is separated from the blood by a selectively permeable membrane, which lets small molecules through, much like the kidney's natural filter. The concentration of the dialysate is carefully controlled to ensure that only excess salts and water pass into it. It has less water and salts in it than the patient's blood has, and so salts diffuse into the dialysate down a concentration gradient, and water moves out of the blood by osmosis. The dialysate is constantly renewed so that the blood continually loses the salts and water that have built up without a properly functioning kidney.

Figure 12.5 A patient undergoing kidney dialysis.

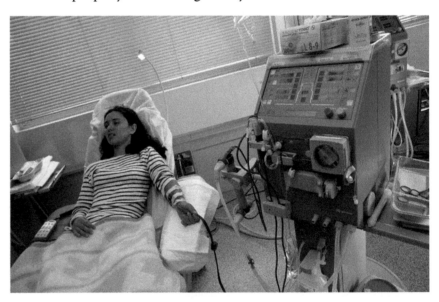

Once the excess salts and water have been removed, the blood is returned to the body.

The problems with dialysis as a treatment for kidney disease are as follows:

▸ The patient normally needs to have dialysis sessions on three days each week, and each session lasts about 4 hours. Some patients have dialysis machines at their homes and do their dialysis while they sleep. However, there are not enough machines to go around, so many patients have to visit hospital three days a week for a session. This is inconvenient and may mean that patients cannot work full-time.
▸ In between sessions, patients have to be very careful about what they eat and drink. They have to restrict fluid and salt intake so that the levels do not build up to dangerous levels, or to an extent where more than four hours of dialysis would be needed.

Kidney transplantation

A kidney transplant is a cure for kidney disease, and patients can live a normal life without having to repeatedly undergo dialysis. However, there are some disadvantages:

▶ The process involves surgery, which always carries some risk.

▶ The 'foreign' kidney transplanted will be recognised by the body's immune system, which will then attack it. To avoid this rejection, kidney transplant patients have to take drugs that suppress their immune system (often for the rest of their lives). These drugs are called immunosuppressants. Their use can make patients more likely to get infections.

▶ Transplanted kidneys have a limited lifespan. Only about 40–50% of transplanted kidneys last longer than 15 years. A young patient may therefore need several transplants during their lifetime.

▶ A person needs to be reasonably strong and healthy (other than suffering from kidney disease) in order to have a transplant. Patients who are very sick are not able to have the operation as the risks are too great.

▶ A person can only receive a kidney from someone who has a similar 'tissue type'. Some patients have to wait many years for a suitable donor.

Chapter summary

- The kidneys regulate the water content of the blood and remove waste products such as urea from the blood. This is necessary because urea is a poisonous waste, and if the blood is too concentrated or too dilute it can disrupt processes in the body.
- The human excretory system contains the following structures: kidneys, renal arteries, renal veins, aorta, vena cava, ureters, bladder and urethra.
- A kidney has the following parts: cortex, medulla, pelvis and ureter.
- A nephron consists of the following structures: capillary knot, Bowman's capsule, tubule, collecting duct, capillary network, and arterioles to and from the capillary knot.
- Blood is filtered under pressure in the capillary knot. The pressure results from the arteriole leaving the knot being smaller than the one entering.
- Glucose, some salts and much of the water are selectively reabsorbed as the fluid moves through the kidney tubule.
- Urine – containing urea, water and excess salts – passes from the kidneys in the ureters to the bladder, where it is stored before being passed out of the body.

- The presence of blood or cells in the urine indicates disease in the kidney.
- The kidneys produce dilute urine if there is too much water in the blood or concentrated urine if there is a shortage of water in the blood.
- The regulation of water absorption in the kidney is controlled by anti-diuretic hormone (ADH).
- Dialysis and transplantation can be used to treat kidney failure.
- In a dialysis machine excess salts and water pass from the blood into a fluid called the dialysate by diffusion and osmosis. The concentration of the dialysate is carefully controlled to regulate this.
- Kidney transplants require a donor with a similar 'tissue type' to the recipient.
- The donor kidney may be rejected by the body and attacked by the immune system, unless drugs are taken that suppress the immune response.
- Dialysis and transplantation each have advantages and disadvantages.

▶ Chapter review questions

1 The diagram shows the excretory system of the human body.

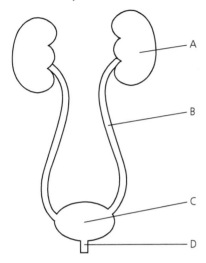

a) From the diagram, copy and complete the table below. [3]

Letter on diagram	Name of structure	Function
	ureter	
		carries urine out of the body
C		

b) Name two waste substances excreted in urine. [1]

c) State how the concentration of the urine changes when there is too little water in the blood. [1]

(from WJEC Paper B3(F), Summer 2015, question 4)

2 The diagram shows some of the processes that control the composition of blood and urine.

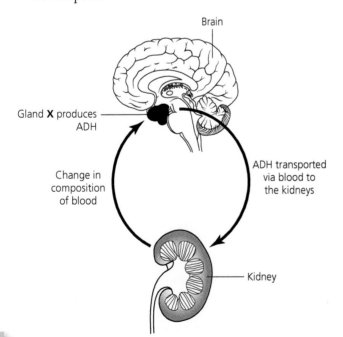

a) Identify the stimulus that causes gland **X** to release ADH. [1]

b) Describe the effect of an increase in ADH production on the kidney and on the composition of urine. [3]

(from WJEC Paper B3(H), Summer 2015, question 7)

3 Read the passage below and answer the questions that follow.

Patients with kidney failure either have a kidney transplant or are treated by dialysis. Dialysis usually involves being attached to a kidney machine for around 4 hours, usually 3 days a week. Most kidney patients have to visit hospital for their dialysis.

Another form of dialysis is peritoneal dialysis, which uses the membrane lining the inside of your abdomen (the peritoneum) as the filter, rather than a machine. Like the kidneys, the peritoneum contains thousands of tiny blood vessels, making it a useful filtering device. An incision is made near the belly button and a thin tube called a catheter is inserted into the space inside the abdomen. This is left in place permanently. A sterile solution containing glucose and salts is pumped through the catheter. Waste products and excess water is drawn out of the blood capillaries lining the peritoneum. The used fluid is drained into a bag a few hours later and replaced with fresh fluid. The fluid needs to be changed around 4 times a day, although some patients use a machine that can exchange the fluid as they sleep. The dialysis can be done at home.

a) What is the main waste product that needs to be removed from the blood? [1]

b) Why does the fact that the peritoneum 'contains thousands of blood vessels' make it a useful filtering device? [1]

c) Why does the fluid need to be sterile? [1]

d) What process will take water from the blood vessels into the abdominal cavity? [1]

e) Suggest two reasons why glucose solution is used rather than water. [2]

f) From the information in the passage, give one advantage of peritoneal dialysis compared to normal dialysis using a kidney machine. [1]

13 Microorganisms and disease

▶ Microorganisms – good or bad?

The term microorganism, as the name suggests, is used to describe any living organism that is microscopic – that is, it can only be seen with a microscope. Examples include viruses, bacteria, microscopic fungi and protists. Microorganisms exist in huge numbers and in a huge variety – there are about 10 million trillion microorganisms for each human on the planet! It is not surprising, therefore, to find out that some are 'good', some are 'bad' and most are neither good nor bad. In going about their daily lives, some microorganisms can cause disease and some cause inconvenience, like spoiling our food. We distinguish the ones that cause disease by calling them pathogens, but the vast majority of microorganisms are harmless. Many even perform vital functions. We have bacteria on our skin that help to keep it in good condition, and others in our gut that help out with digestion. Even the spoiling of food is just an unfortunate side effect of a vital function that microorganisms perform, which is breaking down dead cells and organisms so that the nutrients in them can be recycled. It's just bad luck that some of those dead cells and organisms we call 'food'.

Key term

Protist An organism belonging to the Kingdom Protista. Many protists consist of just one cell, which is eukaryotic (that is, contains a nucleus).

▶ What do bacteria and viruses 'look like'?

Bacteria

A bacterium consists of a single cell, but there are a number of differences between them and the cells of animals and plants.

These differences are shown in Figure 13.1. It is thought that the first forms of life were probably bacteria.

Figure 13.1 Features of a bacterial cell.

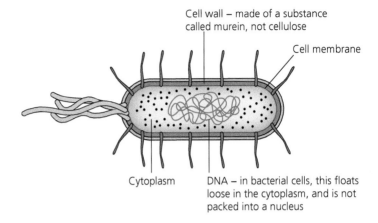

Cell wall – made of a substance called murein, not cellulose

Cell membrane

Cytoplasm

DNA – in bacterial cells, this floats loose in the cytoplasm, and is not packed into a nucleus

Viruses

The structure of a virus is so different from the cells we have seen so far that it cannot really be called a cell at all. The structure of the influenza (flu) virus is shown in Figure 13.2.

A virus is really just some genes in a protein coat. There is no cytoplasm and no cell membrane.

Viruses are even smaller than bacteria, and the first image of a virus was only seen in 1931. Most scientists believe that they are not true living organisms, for the following reasons:

▶ They can be crystallised. This is more characteristic of chemicals than of living organisms.
▶ They can only reproduce by using the resources of a host cell.
▶ They need to be inside a host cell in order to reproduce – they do not have their own 'metabolism'.

On the other hand, they do have genes, and they can reproduce themselves, even if they need to be inside another living cell to do it. Once they have reproduced, the new viruses are released (which destroys the host cell) and infect other cells. One scientist called them 'organisms on the edge of life', which is probably a good description.

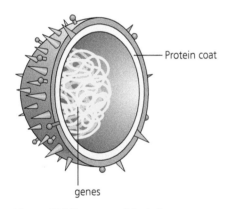

Protein coat

genes

Figure 13.2 Structure of the influenza virus, shown in cross-section.

How can we grow microorganisms?

By definition, microorganisms are very small and difficult to see. In order to study them, scientists grow them into large numbers, which they can work with. This is often done on a special type of jelly called agar, which has nutrients added to feed the microorganisms, in a special plate called a Petri dish. Bacteria grow very quickly indeed, and each bacterium that lands on the agar will soon grow into a circular patch called a colony, which can be seen with the naked eye. Colonies of bacteria are shown in Figure 13.3.

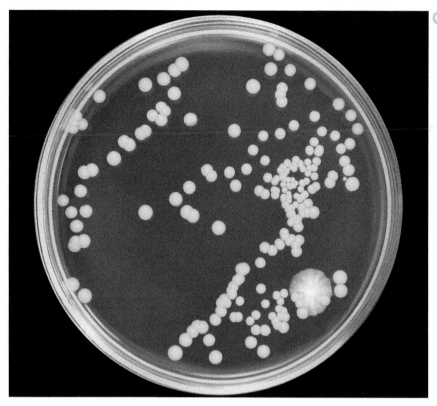

Figure 13.3 Agar plate showing bacterial colonies.

💬 Discussion point

It is recommended that you should never thaw food out from a freezer and then later re-freeze it again. Why?

🧪 Safety notes

- Always make sure to wash your hands thoroughly before and after any microbiology activity.
- Always use a freshly disinfected surface, and remember to disinfect the surface again immediately after the practical.
- Take care when holding the stopper of the culture tube, so that it does not become contaminated.

Because each bacterium grows into a colony, by counting the colonies we know how many bacteria were originally put onto the plate in the original sample.

On average, bacteria divide every 20 minutes, which means that the population grows incredibly quickly. At that rate, one bacterium will have grown into 262 144 in six hours (and remember, that number will continue to double every 20 minutes!). The rate at which bacteria grow is affected by temperature. Warm temperatures are optimum for growth. At low temperatures, the rate of growth is slower and at very high temperatures, the bacteria are killed. We put fresh food in refrigerators because the cool temperature means that the bacteria will grow slowly, and so the food will last longer before going off. To keep it even longer, we put it in a freezer, where the temperature is so cold it virtually stops bacterial growth altogether. Freezing food does not kill the bacteria, however – it just stops them growing.

Aseptic technique

When growing microorganisms, scientists often want to grow just a single species, or find what microorganisms are present in a specific sample. The air is full of bacteria, and it is important that those bacteria do not contaminate the agar plate. To do this, scientists use aseptic technique. This technique is shown in Figure 13.4. Notice how precautions are taken to prevent contamination at each stage.

1.

The wire loop is placed in the Bunsen flame. Kills bacteria already on the loop so they will not contaminate the plate.

2.

The stopper is removed from the bacterial culture tube.

3.

The mouth of the tube is 'flamed'. Ensures that bacteria drifting out of the culture tube are killed.

4.

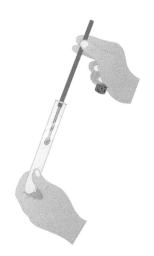

The loop is dipped into the culture to pick up bacteria.

5.

The mouth of the tube is 'flamed' again and the stopper replaced. Kills bacteria that may be drifting into the culture tube, to stop that getting contaminated.

6.

The lid of the Petri dish is partly raised and the loop is drawn across the agar surface. The plate is not completely uncovered, so bacteria from the air will not be able to fall onto it.

7.

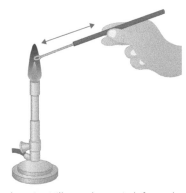

The wire loop is flamed again. Kills any bacteria left on the loop so they will not escape into the environment. Note that sometimes a loop is not used and a sample of the culture is simply poured onto the plate. In such cases, steps 1, 4 and 7 are left out, and steps 5 and 6 are reversed.

Figure 13.4 Aseptic technique.

Specified practical

Investigation into the effect of antibiotics on bacterial growth

An antibiotic is a chemical that kills or inhibits the growth of bacteria. Hand sanitiser gels contain antibiotics, and this experiment compares the effectiveness of some of these products.

Apparatus

> culture of *Staphylococcus albus* on nutrient agar in a Petri dish
> 3 different hand sanitiser gels, each in a 5 cm³ syringe
> liquid soap, in a 5 cm³ syringe
> cork borer
> mounted needle
> marker pen
> sticky tape
> access to incubator at 30 °C
> access to disinfectant to wipe down bench once the plates have been prepared

Procedure

1 Cut four wells in the agar as shown in Figure 13.5, using the cork borer. The agar can be removed from the well using the mounted needle.
2 Label the wells 1 to 4 on the underside of the Petri dish using the marker pen.
3 Fill each well with the different types of hand sanitiser gel (wells 1–3) and the liquid soap (well 4) (Figure 13.6).
4 Seal the lid onto the Petri dish, using sticky tape. Wrap two pieces of tape around the dish, at right angles to one another.
5 Incubate the dish for several days at 30 °C.
6 After incubation, examine the dish for growth of the bacteria. Where the gel has stopped growth, there will be a clear zone around the well. Measure the diameter of this clear zone (Figure 13.7).

Safety

> When handling the Petri dish, avoid leaving the agar surface exposed to the air.
> When filling the wells, keep the agar partially covered by the lid of the Petri dish as shown.
> The incubation temperature should be no higher than 30 °C, to avoid growing microorganisms that could infect humans. (Such microorganisms are adapted to grow at 37 °C, human body temperature.) Room temperature could be used for incubation.
> Wash hands immediately before handling the agar plate and again after the procedure.

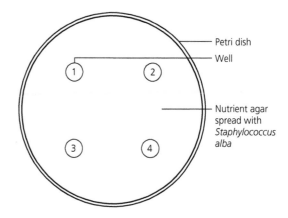

Figure 13.5 Wells cut into the agar.

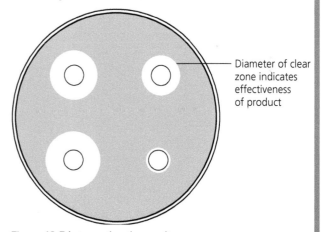

Figure 13.6 Method of filling wells, keeping agar partially covered by the Petri dish lid.

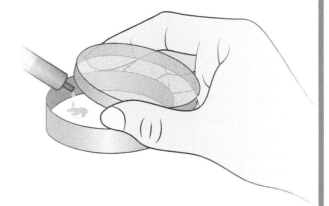

Figure 13.7 Interpreting the results.

Questions

1 Which product was the most effective at stopping the growth of the bacteria? Give reasons for your answer.
2 How strong do you think the evidence is for your conclusion given in question 1? Explain how you reached this decision.

✔ |Test yourself

1 How is a bacterial cell wall different from a plant cell wall?
2 What is a pathogen?
3 What feature do viruses have in common with both bacteria and with plant and animal cells?
4 What name is given to the technique used to prevent contamination of cultures of microorganisms?
5 Why is it best to put fresh foods straight into a refrigerator when you get them home?

▶ How is penicillin produced?

Penicillin is an antibiotic that is commonly used to treat bacterial diseases. Like all antibiotics, it has no effect on viruses. It is produced by several species of *Penicillium* (a fungus). The fungus is grown industrially in a tank called a fermenter (Figure 13.8).

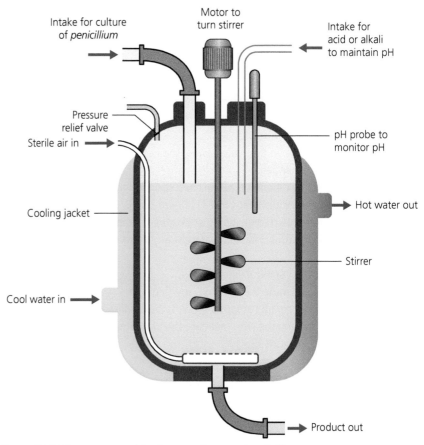

Figure 13.8 A fermenter used for the manufacture of penicillin.

Conditions are controlled in the fermenter to ensure that the fungus has the optimum environment for growth. These conditions are:

▶ **Temperature** – The optimum temperature for growth is about 23–28 °C.
▶ **pH** – Kept at approximately pH6.5.
▶ **Oxygen level** – *Penicillium* is an aerobic fungus so it needs oxygen for respiration and growth.

Although *Penicillium* needs nutrients, these are not added once the growth has started, as the fungus only produces penicillin when the nutrient levels are low. After about 200 hours the liquid is drained from the fermenter, filtered to remove the fungal cells, and then treated chemically to extract the penicillin drug.

How do pathogens spread?

A pathogen is an organism that causes disease. Many of these are microorganisms, mostly viruses, bacteria and fungi. Some pathogens and their associated diseases are shown in Table 13.2.

Table 13.2 Some pathogens and the diseases they cause.

Disease	Caused by	Type of organism
Influenza	Flu virus	Virus
Common cold	Cold virus	Virus
Food poisoning	*Salmonella* or *E. coli*	Bacterium
AIDS	Human immunodeficiency virus (HIV)	Virus
Cholera	*Vibrio cholerae*	Bacterium
Chlamydia	*Chlamydia trachomatis*	Bacterium
Athlete's foot	Dermatophytes (various)	Fungi
Malaria	*Plasmodium* (various)	Protist

Pathogenic species have ways of moving from one host to another and infecting new individuals – that is, they are infectious. There are a number of ways by which pathogens can be spread.

- **Direct contact or body fluids** – Some diseases (for example, skin infections) are passed on by skin-to-skin contact. Personal contact can also pass on diseases carried by body fluids, such as saliva, semen, vaginal fluids and blood.
- **Aerosol infection** – Coughing, sneezing and even talking and breathing can spread disease as they spread tiny droplets containing the pathogen into the air, which can then be breathed in by another person.
- **Water** – If pathogens find their way into water that is then drunk by human beings, the disease can be passed on (Figure 13.9).

Figure 13.9 Drinking contaminated water is likely to spread disease. Note that contaminated water does not necessarily appear 'dirty' as it does here.

This is a particular problem in countries where not everyone has access to piped and treated drinking water.

▶ **Insects** – Insects can act as vectors of disease. The insect may pick up the infection when biting one human being, then transmit it when biting another individual.

▶ **Contaminated food** – Bacteria will feed on most things that we eat. Some of these bacteria will then cause disease when swallowed. The main risk is from food that is starting to go off or that has not been hygienically handled during preparation.

How can we prevent the spread of disease?

HIV/AIDS

AIDS stands for Acquired Immune Deficiency Syndrome and it is caused by a virus – the Human Immunodeficiency Virus (HIV). It is spread by body fluids, mainly blood, semen and vaginal fluids. The virus is found in saliva, too, but in such low numbers that transmission by saliva is virtually impossible – you cannot transmit HIV by kissing, as was once thought. The main means of spread is either sexual intercourse or the sharing of needles in drug use (blood from one user remains on the needle and in the syringe and is then injected into the blood supply of another person).

People with AIDS have a damaged immune system, as HIV infects the white blood cells that play a major role in protecting the body against disease. Immediately after infection, the patient may experience symptoms that are a bit like having flu. These disappear and there may be no further symptoms for many years, but during this time the immune system is being progressively damaged. Eventually, it gets to such a poor state that other infections take hold, and this can be fatal.

The spread of HIV can be prevented by the use of condoms during sexual intercourse, avoiding sharing needles and syringes and wearing surgical gloves when treating bleeding.

Chlamydia

Chlamydia is a sexually transmitted disease caused by a bacterium, *Chlamydia trachomatis*. It is transmitted during unprotected sex, and in many people is symptomless. However, it can cause pain when urinating, unusual discharges from the penis or vagina, painful testicles and bleeding between periods. It can be treated with antibiotics but if left untreated can lead to long-term health problems and sterility.

Using a condom during sexual intercourse prevents the transmission of the disease.

Malaria

Malaria is a tropical disease that kills a huge number of people each year. In 2012, for example, according to the World Health Organization, there were 207 million cases worldwide, and 627 000 deaths. The disease is caused by a single-celled parasite of the species *Plasmodium*, which is carried by mosquitos. The parasite is in the red blood cells that are picked up by the mosquito when it bites. It has no effect on the mosquito vector but can be injected into another person's bloodstream when an infected mosquito bites again. The symptoms include a high temperature, sweats and chills, headaches, vomiting, muscle pains and diarrhoea. The disease can be fatal. In some people, malaria can recur years after the first batch of symptoms. The parasite normally lives in red blood cells but can also travel to the liver, where it can remain dormant for years before re-infecting the blood cells.

The methods of preventing the spread of malaria mostly focus on the mosquito rather than the parasite itself. Methods include:

▶ using mosquito nets at night (when mosquitos are active) and insect-repellent creams and lotions (Figure 13.10)
▶ draining swampy areas where mosquitos breed
▶ treating homes with insecticide to kill the mosquitos
▶ taking antimalarial tablets when travelling to infected areas, but these tablets are only effective while they are being taken – they provide no long-term immunity and so are not practical to use with the populations of infected areas.

Figure 13.10 Mosquito nets around beds prevent people getting bitten by mosquitos while they are asleep.

✔ Test yourself

6 In a penicillin fermenter, why is the air sterilised before it enters the fermenter?
7 A penicillin fermenter is cooled with water. Why is it likely to get hot?
8 Insects can be 'vectors' of disease. What does that mean?
9 What is the difference between HIV and AIDS?
10 What part of the human body is infected by the malarial parasite?

▶ How does the immune system defend against disease?

The human body defends itself against disease with two lines of defence. First, there are features that stop pathogens entering the body in the first place. If that line is breached, then the immune system of the body kicks into action to kill any pathogens that enter.

Human skin is an impenetrable barrier against microorganisms, which covers nearly all of the body. It works very well at stopping the entry of microorganisms as long as it is intact. If the skin is damaged by cuts or burns, then the blood clots and plugs the gap, once again sealing off the entrance to the body while the skin heals and restores the barrier. Blood clotting cannot be quick enough to stop the entry of microorganisms completely, however. Pathogens can also enter through the body openings, which have no skin.

Once pathogens get into the body, white blood cells work in various ways to kill them. The blood is an ideal place to house our immune system, as it penetrates all parts of the body and can reach infections wherever they occur. Two types of white blood cells attack invading microorganisms. Phagocytes ingest ('eat') microorganisms and digest them. In addition, other white cells called lymphocytes produce chemicals called antibodies, which destroy microorganisms, and antitoxins, which neutralise any poisons produced by the pathogens. (One reason why pathogens cause disease is because they produce waste chemicals that are poisonous to human cells.) The two types of white blood cell are shown in Figure 13.11.

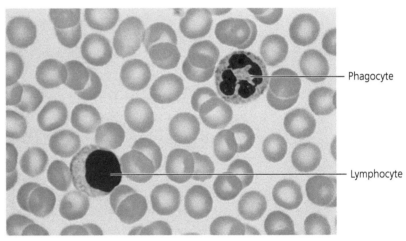

Figure 13.11 A blood 'smear' showing a phagocyte and a lymphocyte.

Antigens and antibodies

In order to attack invading microorganisms, the immune system has to identify them as 'foreign' and so needs to distinguish them from the body's own cells. The way this is done relies on the fact that all cells have patterns of molecules on their surface, and the pattern is different for each individual. These molecules are called antigens, and all the cells of your body have identical antigens. If the white blood cells come across any cell that does not have the 'correct' pattern of antigens, they know it is an invader, and attack it.

The lymphocytes respond to foreign antigens by producing chemicals called antibodies. Phagocytes will attack any foreign cell, but the lymphocytes' response is specific – the antibodies they produce will depend on the antigens detected. The antibodies can either destroy the microorganisms, or may stick them together so that the phagocytes can ingest a lot at once.

When the body encounters a new pathogen that it has not met before, there are no specific antibodies for its antigens. The phagocytes still attack, but the lymphocytes take a while to develop the proper antibodies. During this time, the pathogen may reach levels that cause symptoms of the disease. Once the lymphocytes have made the antibodies, however, they form memory cells, which 'know' the right antibody for the disease. If the same pathogen is encountered again, these memory cells produce the appropriate antibodies very quickly, and the pathogen is wiped out before the infection takes hold. That is why, with many diseases, getting it once makes us immune to it from then on.

💬 Discussion point

No matter how many times we get colds and flu, we never get immune to those diseases. Why not?

▶ How does vaccination work?

Although we can naturally become immune to a disease once we've had it once, for more serious diseases it would be better never to have the disease at all. We can avoid a disease and still get immunity to by having a vaccination against it. Vaccines can protect against both bacterial and viral diseases.

When you are vaccinated against a disease, you are actually injected with the microorganisms that cause the disease. However, the pathogen has either been killed or (less often) weakened so that it is incapable of causing symptoms. It still has its antigens, so the lymphocytes can react to it and build up memory cells and therefore the body becomes immune to that disease. The microorganisms do not reproduce inside the body, and so with a vaccine the immune response is not as great as it is when you get the disease. In order to build up enough memory cells to give full immunity, one or more 'booster' injections are sometimes needed after the first vaccination.

It makes sense to give people immunity to serious diseases as soon as possible, as it is impossible to tell when they will encounter a given pathogen. So, most vaccinations are given to children. The decision to have the vaccination is therefore made by the parents. This is not an easy decision, for a number of reasons.

▶ The vaccination often involves an injection, which can hurt and can scare young children.
▶ Vaccinations usually have some side effects. These are minor (for example, itching and inflammation of the injection site or feeling unwell for a day or so) but in the past some scare stories have put parents off the idea of vaccinations.

In 1998 a small and scientifically unsound study on the MMR vaccine (against measles, mumps and rubella) suggested a link with autism, a disorder that affects the nervous system. Many parents were worried and decided not to let their children have the vaccine, with the result that the incidence of measles (a serious and sometimes fatal disease) grew greatly over the next 15 years. By 2012, however, vaccination rates had recovered to pre-1998 levels and so it is likely that the measles rates will now decline again.

✔ Test yourself

11 What is the name of the type of white blood cell that produces antibodies?
12 What is an antigen?
13 Why do you often become immune to a disease after having it once?
14 Why don't the microorganisms in a vaccine give you the disease?
15 Why might you need booster vaccinations to make you fully immune to a disease?

▶ What are 'Superbugs'?

Over the course of history, scientists have discovered chemicals in living things that have anti-bacterial properties – either killing bacteria or preventing their growth. Once isolated and purified, these **antibiotics** have been used as medicines to supplement the body's own defences. (Viruses are unaffected by all antibiotics – they live inside cells and so the antibiotics cannot reach them.)

Recently, however, certain species of bacteria have evolved resistance to antibiotics. Doctors now use a large variety of antibiotics and resistance to one or a few of them would pose no major problem, but some bacteria have evolved resistance to most of the antibiotics used. The media have called these bacteria 'superbugs', and one that has received a lot of attention is MRSA (methicillin-resistant *Staphylococcus aureus*; Figure 13.12). *Staphylococcus aureus* is a common bacterium, usually carried on the skin, where it can sometimes cause boils or mild skin infections. If it gets through the skin it can cause life-threatening conditions such as blood poisoning.

Natural selection in the bacteria has caused the evolution of antibiotic resistance, as described in Chapter 10. The rate of natural selection has been increased by the extensive use of antibiotics and this could cause a major problem. There are still a few antibiotics that can treat MRSA infections, but it seems that the bacterium is evolving resistance faster than we can produce new antibiotics. MRSA is particularly dangerous in hospitals, where people are already unwell or have wounds from accidents or surgery. The control measures for MRSA fall into two categories – the prevention of infection and combating the evolution of resistance.

To prevent infection, the following measures are adopted:

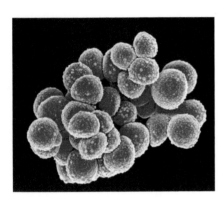

Figure 13.12 Methicillin-resistant *Staphylococcus aureus* bacteria, MRSA.

- ▶ Patients entering hospital are screened for MRSA.
- ▶ Hospital staff wash their hands after going to the toilet, before and after eating and before handling patients in any way. This sort of hygiene should also be practised by the general public, too.
- ▶ Visitors are recommended to wash their hands or use hand sanitiser gels when entering wards.
- ▶ Stringent hygiene measures are taken with any procedure involving body openings or wounds.

To slow natural selection, the following measures are adopted:

- ▶ Doctors avoid prescribing antibiotics wherever possible – for example, if an infection is mild and the body can overcome it without antibiotics.
- ▶ Where antibiotics are prescribed, doctors vary the type as much as possible. Extensive use of any single antibiotic increases the risk that it will become ineffective.

The effect of aspirin on amylase enzyme

One common side effect of drugs is to inhibit the activity of enzymes that may be important in the body. This experiment looks at whether aspirin affects the activity of a given enzyme.

The enzyme used is amylase. In the body amylase catalyses the conversion of starch into maltose. One of the places it is found is in saliva, which starts the breakdown of starch in food. The activity of the enzyme can be tested with iodine in potassium iodide solution. Iodine turns blue-black in the presence of starch, but remains orange-brown with maltose.

Apparatus

> paper cup, containing 40 cm³ tap water
> 40 cm³ 2% starch suspension
> 5 cm³ ethanol
> 5 cm³ 0.5% salicylic acid (aspirin) solution
> 0.01 M iodine in potassium iodide solution

> 4 test tubes
> test-tube rack
> 3 × 5 cm³ plastic syringes
> 1 cm³ plastic syringe
> stopwatch
> glass-marking pen
> eye protection

Hypothesis

Aspirin inhibits the action of amylase enzyme.

Safety notes

> Wear eye protection.
> You will be supplied with a risk assessment by your teacher.

Procedure

1 Number the test tubes from 1 to 4.
2 Using a 5 cm³ syringe, put 5 cm³ starch suspension into each tube.
3 Fill a clean 5 cm³ syringe with salicylic acid solution and add 0.5 cm³ to tube 2, 1 cm³ to tube 3 and 2 cm³ to tube 4.
4 Take water from the paper cup into your mouth. Swill it around thoroughly to get a sample of saliva rich in amylase, then spit it out into the cup.
5 Using a clean 5 cm³ syringe, add 2 cm³ saliva solution to each tube and gently rotate the tubes to mix their contents. Start the stopwatch.
6 After 10 minutes, use the 1 cm³ syringe to add 0.5 cm³ iodine in potassium iodide solution to each tube (Figure 13.13). If the colours are faint add a further 0.5 cm³ iodine solution to each tube. Estimate the colour intensities of each tube using the scale in Table 13.3. Zero on the scale means the starch

has been unaffected by the amylase; 4 on the scale means the starch has been completely converted to maltose by the amylase.
7 Record your results in a suitable table.
8 Display your results in a graph.

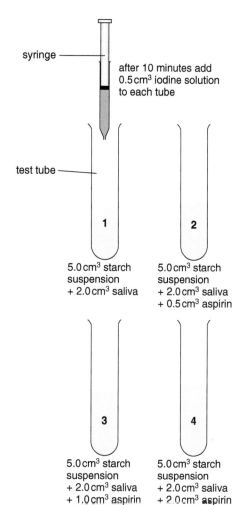

Figure 13.13 Apparatus for an experiment investigating the effect of aspirin on amylase enzyme.

Table 13.3 Reference scale used to assess the extent to which starch has been digested by amylase.

Colour	Unit scale
Brown, pale	4
Brown, dark	3
Brown-purple, pale	2
Brown-purple, dark	1
Blue	0

Analysing your results

To what extent do your results support the stated hypothesis 'Aspirin inhibits the action of amylase enzyme'?

► How are medicinal drugs tested?

Antibiotics are just one type of a variety of drugs that have been developed for medicinal use. A drug is a chemical that alters the way the body works in a certain way. The main effect is usually the one that is needed and the reason that the drug is taken. However, drugs often have other effects that are not needed, called **side effects**. These may be minor or serious. If there are serious side effects from taking a drug, a decision has to be made as to whether the benefit from the drug is worth the side effect. If not, the drug will not be released for use.

When a potential new drug is discovered, it goes through a rigorous and lengthy test procedure before it is released for general use. The development time for a new drug can be as much as 20 years before it is released for general use. Once a drug has been isolated in a form in which it can be used, approval involves several stages, each of which must be passed before progressing to the next.

1 The drug is tested on human cells grown outside the body in a laboratory.
2 The drug is then tested on animals, which are monitored for side effects.
3 The drug then undergoes a clinical trial. It is tested on healthy volunteers, initially in very low doses.
4 Further clinical trials are carried out to establish the optimum dose for the drug.
5 The drug is then trialled with a sample of people who have the disease or condition it is intended to treat, to see if it is more effective than current treatments.
6 If all of these tests are passed, the drug is then licensed for general use.

Stages 1–4 are referred to as **pre-clinical testing**. Stage 5 is **clinical testing**.

► Why are monoclonal antibodies important for medicine?

Lymphocytes can be cultured in the laboratory and be induced to produce antibodies. Activated lymphocytes of a specific type are able to divide continually, and produce very large numbers of cells, each of which will produce a single antibody. This allows scientists to obtain large quantities of a specific antibody. These antibodies, which all respond to the same antigen, are called monoclonal antibodies, and they have huge potential for use in medicine.

Monoclonal antibodies are produced by the following procedure, which is shown in Figure 13.14.

1 The antigen that the antibody is to be used to combat is injected into a mouse.
2 The mouse's white blood cells will produce antibodies specific to that antigen (after several days). Many of these white blood cells can be found in the spleen.
3 Some spleen tissue containing white blood cells is collected from the mouse.

4 These cells are fused with **myeloma** cells (cancerous white blood cells) to form **hybridoma** cells. Cancer cells are used because they keep on growing and dividing continually.

5 The hybridoma cells are collected and grown in a culture medium designed to support them, but not any left-over myeloma cells, which die.

6 The hybridoma cells also have the property of growing and dividing indefinitely, and so produce large amounts of (monoclonal) antibody specific to the original antigen.

7 The antibodies are extracted from the culture medium by centrifugation, filtration and chromatography.

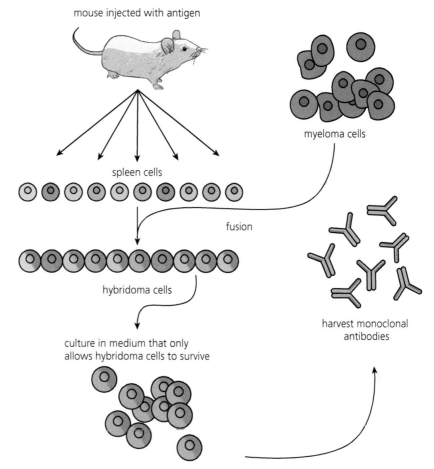

mouse injected with antigen

myeloma cells

spleen cells

fusion

hybridoma cells

culture in medium that only allows hybridoma cells to survive

harvest monoclonal antibodies

Figure 13.14 The process by which monoclonal antibodies are produced.

Monoclonal antibodies will attach to specific antigens and therefore to specific cells or viruses. Scientists can attach other molecules to the antibody. Some of these molecules are detectable (therefore making the cell or virus detectable) while others will kill the targeted cells. Some possible uses are listed here:

▸ Monoclonal antibodies that attach to antigens for the *Chlamydia* bacteria or the HIV virus can be used to diagnose those diseases. The monoclonal antibody has some sort of detectable chemical 'label' attached to it. This may be a radioactive isotope, a fluorescent dye or an enzyme that will cause a colour change of some sort. If a patient's blood displays the detectable label, the antigen must be present.

- Monoclonal antibodies can be used to identify different tissue types, an essential process for matching transplant patients and donors. The monoclonal antibody is designed to recognise antigens on the surface of the patient's cells, and is then used to detect similar antigens in potential donors. If the antigens are detected in a potential donor, his or her tissue will be a match for the patient.
- The malarial parasite has antigens that can be detected using monoclonal antibodies, allowing scientists to diagnose the disease in patients that are not showing symptoms. The antibody is developed to detect antigens which are only found on the malarial parasite, with a detectable chemical label attached. This has helped to track the spread of the disease in certain areas.
- Monoclonal antibodies specific to cancer cells can have a drug attached to them that will kill the cells. This means that chemotherapy will be better targeted, and the drugs used will not damage healthy cells in the way they have done in the past.

→ Activity

Should scientists use animal testing?

Suppose a new drug is discovered that has the potential to cure or treat a human disease. It is likely that the drug will have some side effects, but how serious those might be will not be known. The scientists, at this stage, cannot test the drug on human volunteers (the law will not allow it) because in extreme cases it might kill them. The alternative is to test the drugs on animals, usually mammals, because humans are mammals. In Britain the law says that any new medical drug must be tested on at least two different types of live mammal, and that one must be a large, non-rodent animal.

Many people are against animal testing but the arguments are complex and there is no simple solution. Consider the following points:

> In the UK animal testers have to be licensed. If they mistreat the animals or use them for unnecessary research they may lose their licence. However, 'not mistreating' means being as humane as possible. Sometimes the tests are unpleasant (for example, taking blood samples) but necessary – such tests are not classed as 'mistreating' the animal.

> There are alternatives to animal testing (for example, testing on human cells grown in the laboratory) but these are not suitable in all cases and a body is more complex than isolated cells, so the results may not be relevant.

> Humans and rats are different animals, and protestors say that this means that results using experimental animals are not necessarily relevant to humans. Testers say such differences can be taken into account in their conclusions. It is likely that animal test results are not entirely applicable to humans but provide some useful information.

> Some people believe that it is unethical to use animals because they cannot 'volunteer' for tests and that animal life should be regarded as being just as valuable as human life. If you believe that, then animal testing is clearly wrong, even if it saves human lives.

> In the past, drugs that would not have been licensed for use without animal testing have saved human lives.

> It is likely that animals do not have emotions like humans. They may not feel fear in the way we do, but it is very difficult to be certain of this.

Do some further research into the issue of animal testing and come to a personal conclusion about its value and whether or how it should be continued. Use evidence to justify your opinions, taking care to avoid bias.

✔ | Test yourself

16 Why are hospital visitors asked to wash their hands before entering wards as a precaution against MRSA?

17 What is the name of the process by which MRSA has become resistant to antibiotics?

18 What is the purpose of doing pre-clinical trials of a drug using healthy people?

19 What are monoclonal antibodies?

⬇ | Chapter summary

- Most microorganisms are harmless and many are beneficial; some microorganisms, called pathogens, cause disease.
- Pathogens include bacteria, viruses, protists and fungi.
- The basic structures of a bacterial cell and a virus are described.
- Microorganisms causing disease can be spread by contact, aerosol, body fluids, water, insects and contaminated food.
- HIV/AIDS, chlamydia and malaria are described, including the microorganisms that cause them, their effects on the body and means of prevention.
- The body defends itself from disease by having an intact skin that forms a barrier against microorganisms, blood clots to seal wounds, phagocytes in the blood to ingest microorganisms and lymphocytes to produce antibodies and antitoxins.
- An antigen is a molecule that is recognised by the immune system – foreign antigens trigger a response by lymphocytes, which secrete antibodies specific to the antigens.
- Antibodies can kill the microorganism concerned or assist the phagocytes' ability to engulf them.
- Vaccination can be used to protect humans from infectious disease.
- Certain factors have influenced, and continue to influence, parents' decisions about whether to have children vaccinated or not.
- A vaccine contains antigens derived from a disease-causing organism.
- A vaccine will protect against infection by that organism by stimulating the lymphocytes to produce antibodies to that antigen.

- Vaccines can protect against diseases caused both by bacteria and by viruses.
- Once an antigen has been encountered, memory cells remain in the body and antibodies are produced very quickly if the same antigen is encountered a second time – this memory provides immunity following a natural infection and after vaccination but is specific to one microorganism.
- Antibiotics, including penicillin, were originally medicines produced by living organisms, such as fungi.
- Antibiotics help to cure bacterial disease by killing the infecting bacteria or preventing their growth, but do not kill viruses.
- Some resistant bacteria, such as MRSA, can result from the overuse of antibiotics – methods have been introduced to try to control MRSA.
- Some conditions can be prevented by treatment with drugs or by other therapies.
- New drug treatments may have side effects and so extensive testing is required.
- There are risks, benefits and ethical issues involved in the development of new drug treatments, including the use of animals for testing drugs.
- Development of potential new medicines includes pre-clinical and clinical testing.
- Monoclonal antibodies are produced from activated lymphocytes, which are able to divide continuously – this produces very large numbers of identical antibodies, specific to one antigen.
- Monoclonal antibodies have medical uses including diagnosis of diseases, tissue typing for transplants, monitoring the spread of malaria and supporting chemotherapy for cancers.

► Chapter review questions

1 The series of diagrams below, labelled A–D, shows stages in the aseptic techniques involved in inoculating and plating bacteria from milk samples. The stages are not shown in the correct order.

Stage

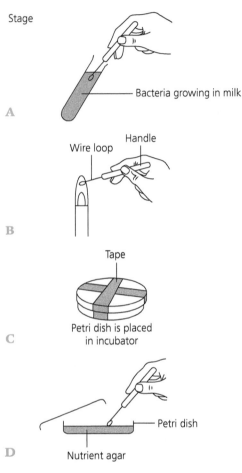

A — Bacteria growing in milk

B — Wire loop, Handle

C — Tape, Petri dish is placed in incubator

D — Petri dish, Nutrient agar

a) State the correct order of the stages.　　[1]

b) Give a reason why the Petri dish is sealed in stage **C**.　　[1]

Students kept fresh pasteurised milk at three different temperatures for five days.

At the end of this time they spread milk samples onto sterile agar plates, which were then incubated at 25 °C. After three days of incubation, the agar plates were examined.

The results obtained are shown below.

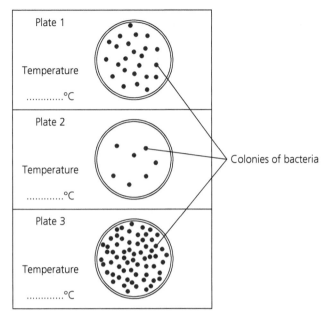

Plate 1 — Temperature ....°C

Plate 2 — Temperature ....°C

Plate 3 — Temperature ....°C

Colonies of bacteria

c) Using temperatures from the list below, give the most likely temperature at which the milk was kept for the five days in each sample before the milk samples were spread onto the agar.　　[3]

> 10 °C

> –10 °C

> 35 °C

> 4 °C

> 150 °C

d) Each of the colonies of bacteria on the agar plates is made up of many thousands of bacteria. How many bacteria were in the original milk sample spread onto plate **2**?　　[1]

e) Explain the possible consequences to this investigation if stage B shown in part (a) of this question had not been carried out.　　[2]

(from WJEC Paper B3(H), Summer 2015, question 3)

2 In 1928, Alexander Fleming found a fungus called *Penicillium* in a Petri dish containing a culture of bacteria growing on agar jelly. The diagram shows what he observed.

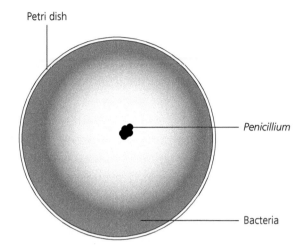

Fleming made two conclusions.

- A chemical released from *Penicillium* was harming the bacteria.

- The chemical was diffusing through the jelly.

a) What is the evidence in the diagram for each conclusion? [2]

The chemical found in *Penicillium* was extracted and is known as penicillin.

b) What name is given to types of drugs such as penicillin? [1]

c) Why has penicillin become less effective at killing bacteria in recent years? [2]

MRSA has become a serious problem in hospitals.

d) Describe one effective control measure used in hospitals against MRSA. [1]

(from WJEC Paper B3(H), Summer 2013, question 5)

3 The antibiotic penicillin is produced in large stainless steel fermenters containing a liquid nutrient culture medium in which *Penicillium* is grown. The diagram shows a fermenter.

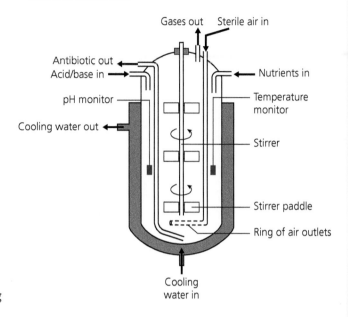

a) Name a nutrient that should be added to the fermenter. [1]

b) Why is air pumped into the fermenter? [1]

c) Why does the air have to be sterile? [1]

d) To which group of living organisms does *Penicillium* belong? [1]

(from WJEC Paper B3(H), January 2013, question 3)

How scientists work

▶ Isn't science all about learning facts?

Science is not just a load of facts that need to be learnt. Science is asking questions about the world around us and trying to come up with answers. Sometimes these answers can be found by careful observation. Sometimes we need to test out a possible answer (a hypothesis) by doing experiments (Figure 14.1). Facts are useful, though. We need to know if someone else has already found the answer that we are looking for (if so, there would be no point in doing an experiment to find it out again, unless we wanted to check the answer was right). Scientific facts may also help us to come up with a hypothesis.

The people we call 'scientists' don't sit around learning facts. They use the facts they already know, or that they can find out by research, to ask questions, suggest answers and design experiments. Science is a process of enquiry, and to be good at science you have to understand and develop certain enquiry skills.

Figure 14.1 Science is all about doing experiments and making observations to find the answers to questions.

▶ What is the scientific method?

The way science is carried out and questions are answered is quite complicated and varied. A flowchart showing the ways in which scientists investigate things, known as the **scientific method**, is shown in Figure 14.2 on the next page. Not every question will involve *all* of these steps. The flowchart shows several skill areas that scientists need to develop – in particular:

- ▶ an ability to ask scientific questions and to suggest hypotheses
- ▶ experimental design skills
- ▶ practical skills in handling apparatus
- ▶ an ability to present data clearly and analyse it accurately (data handling).

This chapter will deal with these skills, which are essential for scientists to master.

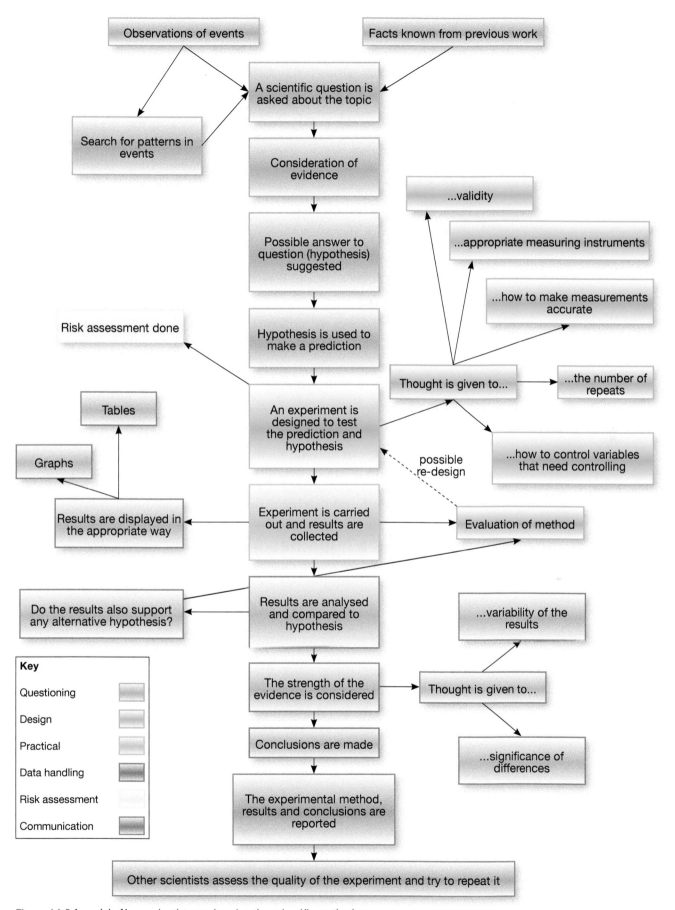

Figure 14.2 A model of how scientists work, using the scientific method.

▶ What is a scientific question?

Sometimes you can ask a question, but there is no hope of getting a definite and undisputed answer. Look at these questions:

- ▶ Is there a God?
- ▶ What would be the best way to spend a lottery win of £10 000 000?
- ▶ Who is the greatest painter that ever lived?
- ▶ Is Paris a nicer place to live than London?

These are not scientific questions. Whether or not there is a God is a matter of faith and cannot be proved in a scientific sense. The other questions are all complex and open to different opinions. Scientific questions have the possibility of being answered by experiments.

Let's consider another question:

- ▶ How can I make the plants in my greenhouse grow better?

This is a scientific question but it's not a very good one. It is possible to answer it by experimenting, but you would have to do a lot of experiments because there are many factors (and combinations of them) that could affect plant growth. A better question would be more specific, such as:

- ▶ What effect does the temperature in my greenhouse have on the growth of the plants inside?

It is possible to find the answer to this by exposing the plants to different temperatures. It would be even better to specify one particular type of plant, because temperature may not have exactly the same effect on all the plants in the greenhouse.

▶ What is a hypothesis?

Sometimes, scientists have an idea about possible answers to a particular question. They look at known facts or make observations and try to explain them, using the available evidence. A suggested explanation is called a hypothesis. A hypothesis is more than just a guess, because it can be justified by scientific evidence and previous knowledge. A hypothesis is not the same as a prediction, but it can be used to make a prediction. A prediction suggests what will happen, but does not explain why, whereas a hypothesis suggests an explanation.

There is no point suggesting a hypothesis if you cannot get any information about whether it is right or wrong, so a scientific hypothesis must be able to be tested by experiment. When scientists do experiments to test a hypothesis, the results can provide evidence that supports or contradicts it. Experiments are generally designed to try to disprove a hypothesis, and sometimes they do. Even if the results do support the hypothesis, they do not *prove* it. If a hypothesis is so well supported by evidence that it is generally accepted, it becomes known as a theory.

In summary, a scientific hypothesis:

- ▶ is a suggested explanation for an observation
- ▶ is based on evidence
- ▶ can be tested by experiment.

→ Activity

Everyday hypotheses

Jane's mum says that she often gets indigestion when she drinks white wine, but not when she drinks red wine. Jane, Aaron, Dave and Rebecca are suggesting hypotheses to explain why (Figure 14.3).

JANE
Wine is acidic and indigestion is caused by too much acid in your stomach. Perhaps white wine is more acidic than red.

DAVE
Wines are not all the same strength of alcohol. Perhaps the white wine is just stronger in alcohol than the red.

Aaron
I think that as people get older they can't drink as much alcohol without showing side effects. Jane's mum is 48.

REBECCA
My mum prefers red wine too. Maybe white wine just upsets your stomach more.

Figure 14.3 Suggested hypotheses for Jane's mum.

Questions

For each person, say:

1 if the suggestion qualifies as a scientific hypothesis
2 if it does, say whether you think it is a good scientific hypothesis.

▶ How do scientists devise a hypothesis?

Scientists have to be able to suggest hypotheses to explain things that they observe, so that they can test them with experiments to find out how and why things happen in the world around them. We have seen in the previous section that there are a number of criteria for a hypothesis.

You actually make hypotheses all the time in everyday life to solve problems. Let's look at an example. You go to use a torch, and find it doesn't work. You immediately make one or more hypotheses.

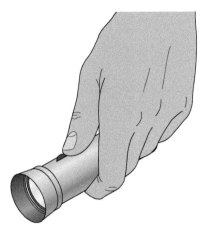

Figure 14.4 Hypotheses for torch not working.

We must now consider the five hypotheses we have thought of (Table 14.1).

Table 14.1 Considering each hypothesis for a torch not working.

	Hypothesis	Evidence	Accept/reject	Can it be tested?
1	Not switched on	Switch is on.	Reject	No need
2	No batteries	Torch was used yesterday and it is unlikely batteries would have been removed since (but not impossible).	Accept	Yes (look to see if there are batteries inside)
3	Flat batteries	Batteries have not been replaced recently and batteries have a limited life.	Accept	Yes (replace batteries)
4	Blown bulb	Bulb has never been replaced but is well within its life span.	Accept	Yes (replace bulb)
5	Poor connection	No evidence either for or against.	Accept	Yes (inspect and clean contacts)

We now have four hypotheses, all of which can be tested. Looking at the strength of the evidence, hypothesis 3 (flat batteries) seems the most likely, and would be easy to test. In looking to replace the batteries, you would also be testing hypothesis 2. If you replace the batteries and the torch still does not light, you would reject hypothesis 3 and go on to test hypothesis 4 or 5.

You do this sort of thing often – you just may not have known that you were developing a hypothesis!

Devising hypotheses

Let's look at an observation and see if you can devise a hypothesis to explain it.

Dogs will often wait by a window or door in their house just before their owner returns home from work (Figure 14.5). This is an observation that must have some explanation if it happens regularly (which dog owners say it does). You need to devise a hypothesis that will explain this behaviour, and fit with any evidence or scientific knowledge.

Let's start by collecting information about the observation. Mark and Anne have a dog called Prince. Anne drives home from work and arrives around 6.00 pm. Mark says that, at around 5.50 pm, Prince sits by the window overlooking their drive and does not move until Anne's car arrives. Prince hardly ever sits by the window at any other time of day.

Evidence and scientific knowledge

> Prince always goes to the window at about 5.50 pm.
> His owner always arrives home about 6.00 pm.
> Prince does not sit by the window at other times of the day.
> Dogs have much more highly attuned senses of smell and hearing compared to humans.

Figure 14.5 This dog is waiting at the window for its owner to arrive.

> All mammals have a **biorhythm** – that is, they are aware of roughly what time of day it is even if they can't read a clock.

Questions

1 Suggest at least two possible hypotheses that might explain Prince's behaviour.
2 Pick one of your hypotheses and suggest how you might test it.

▶ How do scientists design an experiment?

A good experiment will provide an answer to your question, or at least give information that will get you nearer to an answer. If you have a hypothesis, it will provide evidence as to whether the hypothesis is right or wrong (although it may not actually *prove* the hypothesis). We call experiments like this valid experiments. If the experiment design has any major faults, it will probably not be valid.

Two of the most important things that make an experiment valid are fairness and accuracy. If it's a fair test, and your results are accurate, then you are more likely to get the 'correct' answer.

Fair testing

Imagine that you wish to test whether the amount of water added to seeds affects the success of germination. Think of all the variables (apart from the amount of water) that could affect the rate of germination:

- ▶ the type of seed
- ▶ temperature
- ▶ mineral content of the soil
- ▶ light intensity
- ▶ humidity of the air (because that affects how much water evaporates)
- ▶ how close the seeds are to one another.

For the test to be fair, you have to try to ensure that none of these things affects the experiment. You would use different amounts of water added to the seed, because that is the variable you are testing. You would use the same type of seed in all tests, control the temperature (for example, using a thermostatically controlled incubator), use the same soil, have the same light intensity, try to control the humidity of the air if possible, and have the same number of seeds in each seed tray and space them the same way.

Controlling temperature, light intensity and humidity is quite complicated. It would be possible to just leave all the trays in the same place in the laboratory. Room temperature, light intensity and humidity would vary, but in exactly the same way for each seed tray, so the test would still be fair.

Sometimes, in an experiment or a scientific study, there is a variable that you cannot control. In the example above, one or two seeds in a sample may be dead and incapable of germinating, but you cannot tell that. You just have to keep it in mind and take account of it in your analysis. For example, dead seeds would have a small effect on your results. So, for example, if one level of watering produced 85% germination and another produced 80%, you might not count those as 'different' because the number of live seeds could account for the difference. If one tray had 85% germination and another had 60%, then there would really be a difference, as the number of live seeds would probably not vary that much.

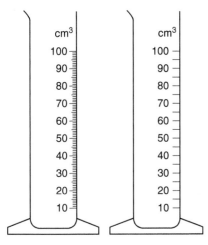

Figure 14.6 The measuring cylinder on the left has a higher resolution (that is, smaller divisions) than the one on the right.

Figure 14.7 This photo shows how much bubbles vary in size. Therefore, 'one bubble' cannot be an accurate measurement of a volume of gas.

Making measurements accurate

Accurate measurements are defined as those that are as near as possible to the 'true' value. The problem is, we don't know exactly what the true value is! So, it is impossible to be certain that a measurement is accurate. All we can do is make sure there are not any obvious inaccuracies.

Any instrument used for measurement should be as accurate as possible. It is usually best to use an instrument with a high resolution (Figure 14.6).

Inaccuracy can result from the units of measurement being imprecise. When measuring the amount of gas given off by a photosynthesising aquatic plant, for instance, counting bubbles is going to be inaccurate because the bubbles will not all be the same size, so that 25 bubbles in one case may contain more gas than 30 bubbles in another, if the first set of bubbles contains more big bubbles (Figure 14.7).

Inaccuracy can also result from human error caused by the means of measurement. If you are timing a colour change, for instance, it is often difficult to judge *exactly* when the colour changes, because it will be a gradual process.

Most measurements are not 100% accurate. This is acceptable as long as the inaccuracy is not so large that it makes comparisons of different measurements invalid.

In the 'counting bubbles' scenario above, for instance, if you have two readings of 86 bubbles and 43 bubbles then, even though there is inaccuracy, the difference is so large that the inaccuracy doesn't matter. If you had two readings of 27 and 32 bubbles, though, you could not confidently say that the second reading was actually larger than the first.

Why do scientists repeat experiments?

Scientists repeat experiments (or take large samples) for the following reasons:

▶ The more repeats you do (up to a point) or the bigger your sample, the more reliable any calculated mean becomes. Note that the individual results do not become more reliable, only the mean.
▶ Repeats or larger samples allow you to identify anomalous results more accurately.

Let's use an example. Suppose you are looking at response time, where a subject has to press a button when a certain signal is seen on a screen. The results are shown on the next page in Table 14.2.

If only three repeats had been done, it might be assumed that the result of trial 2 was anomalous. However, it is obvious when doing repeats that actually trials 1 and 3 were the anomalous ones. The mean after three repeats turns out to be very inaccurate. After 15 repeats the mean has come down considerably and the effect of the two high values in trials 1 and 3 has reduced (though the mean is still a bit high and even more repeats would still be beneficial).

Table 14.2 Results of a response time study.

Trial	Response time, in millisecs	Mean response time, in millisecs
1	536	536
2	240	388
3	498	425
4	258	383
5	260	358
6	248	340
7	236	325
8	302	322
9	233	312
10	241	305
11	245	300
12	256	296
13	233	291
14	250	288
15	241	285

How many repeats?

Scientific results always vary to some extent, and sometimes they vary a lot. This is referred to as repeatability. If the **repeatability** is very good and the results are all close together, then an accurate mean is obtained very quickly and few repeats are needed. If the repeatability is poor, though, you will only be able to have confidence in the mean after many more repeats. It is not unusual for scientists to repeat experiments 30–50 times. In studies involving a sample, a good sample size is usually around 100 (more if a large population is being sampled). A sample size of less than 30 is generally considered statistically invalid.

✔ Test yourself

1 You notice that woodlice tend to gather under logs. Which of the following is the most scientific question to ask?
   a) Why do woodlice gather under logs?
   b) Do woodlice move away from light and so gather in dark places?
   c) Do woodlice actually have preferences about where they go?
   d) Do woodlice like wood?
2 What is the difference between a hypothesis and a prediction?
3 If you do an experiment that requires you to record the time taken for a colour change from red to blue, the results are unlikely to be absolutely accurate. Why?
4 Why is keeping a set of experiments at room temperature an acceptable (even if not ideal) way of controlling temperature?
5 An experiment is repeated three times. Why is this unlikely to be enough?

Experiments with no hypothesis

Not all experiments have a hypothesis. Some are done just to get information. For example, scientists investigating how populations of yeast cells grow over time simply set up a population and counted the cells at different times. They did not have a hypothesis for what was going to happen.

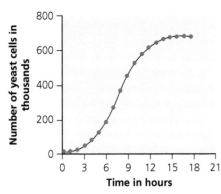

Figure 14.8 Population of yeast cells over time.

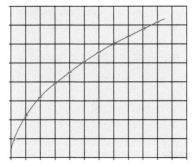
Bar chart

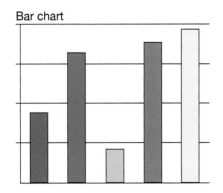
Line graph

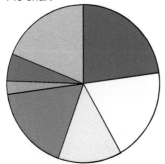
Pie chart

Figure 14.9 Data can be displayed in a variety of ways.

When they got the results shown in Figure 14.8, they then had to think of a hypothesis to explain the curve.

Presenting results

Tables

When scientists record their results, it is important that this is done in a way that is clear to anyone reading their report. Generally, results are presented in tables and usually in some form of graph or chart.

Tables are ways of organising data so that they are clear and the reader does not have to search for them in the text. If you need to look at the method to see what the table means, the table is not doing its job.

▶ Tables must have clear headings.
▶ If the measurements have units, these should be shown in the column headings.
▶ Tables must have a logical sequence to the rows and columns.

Graphs and charts

There are several types of graphs and charts, but the three types most often used are bar charts, line graphs and pie charts (Figure 14.9).

▶ **Bar charts** are used when the *x*-axis shows a **discontinuous variable** (no intermediate values) – for example, months of the year, eye colour, and so on.
▶ **Line graphs** are drawn when the *x*-axis shows a **continuous variable** (any value is possible) – for example, time, pH, concentration, and so on.
▶ **Pie charts** are used to show the make-up of something. Each section represents a percentage of the whole.

Bar charts and line graphs are drawn to show patterns or trends more clearly than a table would. Once again, the graph should display everything necessary to identify the trend, without any need to read through the method.

A good quality bar chart or line graph must have the following:

▶ a title
▶ both axes clearly labelled, with units where appropriate
▶ a 'sensible' and easy-to-read scale for each axis
▶ use of as much of the space available as possible for the scale (without making it awkward to read)
▶ axes the correct way round – if one factor is a 'cause' and the other an 'effect' then the cause (the **independent variable**) should be on the *x*-axis and the effect (the **dependent variable**) should be on the *y*-axis (sometimes, the relationship is not 'cause and effect' and the axes can be either way round)
▶ accurately plotted data
▶ clearly distinguished sets, if more than one set of data is plotted, plus a key to show which set is which
▶ in a line graph, if the data follow a clear trend, this should be indicated with a **line of best fit** – if there is no clear trend, the points should be joined by straight lines, or left un-joined.

▶ How do scientists analyse results and draw conclusions?

Results are usually analysed for one of three purposes:

- ▶ to identify relationships between two or more factors
- ▶ to decide if a hypothesis is likely to be correct
- ▶ to help to create a hypothesis.

Relationships

Relationships are most clearly shown by line graphs. The direction in which the line slopes (or does not) indicates the type of relationship (Figure 14.10). There may be two or more different types of slope on some graphs.

- ▶ When the line slopes upwards, as in Figure 14.10(a), this indicates that as A increases, so does B. This is called a **positive correlation**.
- ▶ When the line slopes downwards, as in Figure 14.10(b), it shows that when A increases, B decreases. This is called a **negative correlation.**
- ▶ If the line is horizontal, as in Figure 14.10(c), it means that values of B are unrelated to A, and there is **no correlation** between the variables.
- ▶ If the graph forms a straight line and it goes through the origin, as in Figure 14.10(d), this is called a **proportional relationship**.

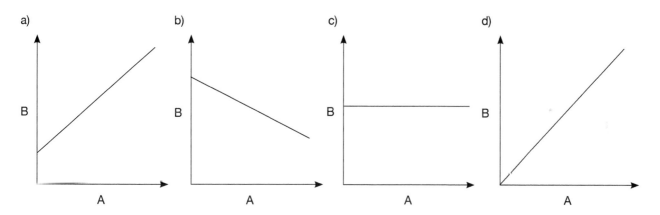

Figure 14.10 Line graphs can show different relationships between the variables.

Just because two factors may show a relationship, it does not mean that either of them actually *causes* that relationship. If B goes up when A increases, it does not mean that the increase in A is what *makes* B go up.

Testing a hypothesis

If there is a hypothesis, the purpose of the experiment is to test it, so the conclusions can only be one of three:

- ▶ The evidence supports the hypothesis.
- ▶ The evidence does not support the hypothesis.
- ▶ The evidence is not conclusive either way.

Experiment can very rarely *prove* a hypothesis.

Centuries ago, the people of Europe believed that swans were always white, because every swan they had ever seen was white. They had the hypothesis that 'all swans are white'. In 1697, though, explorers in Australia found black swans (these have since been introduced into Britain). This instantly disproved the hypothesis, because there could be no doubt whatsoever about the evidence (Figure 14.11). No matter how many white swans the Europeans saw, this would never have proved that all swans are white. Even if black swans had never been discovered, no-one could be certain that there wasn't one somewhere in the world still waiting to be found!

If a long series of experiments have been carried out and all of them support the hypothesis, then scientists treat the hypothesis as if it was true (it becomes a **theory**) even though they still would not say it had been *proved*.

In order to decide whether to continue to accept the hypothesis or reject it, the strength of the evidence is very important.

The flowchart shown in Figure 14.12 shows how scientists arrive at conclusions about a hypothesis.

Figure 14.11 This black swan clearly disproves the hypothesis that 'all swans are white'.

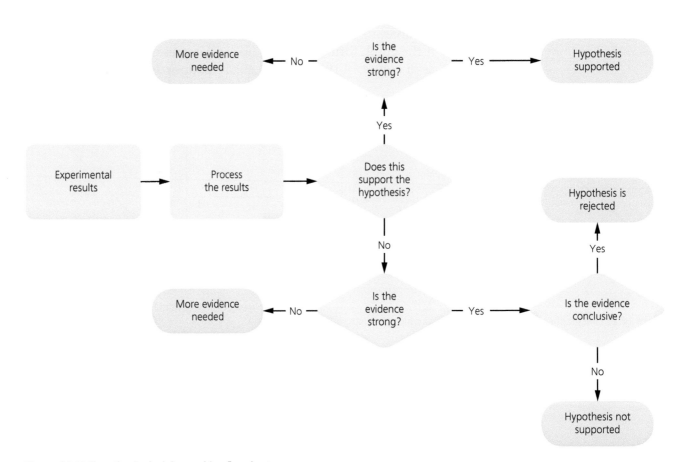

Figure 14.12 Hypothesis decision-making flowchart.

Testing hypotheses

1 Natalie had a hypothesis that wet paper could hold less weight than dry paper. She tested paper bags, adding weight 10 g at a time until the bag broke. She tested 10 bags, and then soaked 10 similar bags in water and tested them. In every single case, the wet bags broke with less weight in them than the dry bags. What should Natalie's conclusion be?

A Her hypothesis is proved.

B Her hypothesis is supported.

C Her hypothesis is doubtful.

D Her hypothesis should be rejected.

2 Glyn had a hypothesis that a certain brand of insulated mug did not actually keep drinks any warmer than a normal ceramic mug. He timed how long it took water to cool by 10 °C in the two types of mug. He ran the test 50 times. On average, the water took 6 minutes longer to cool down in the insulated mug, and in all 50 tests the water in the ceramic mug cooled quicker. What should Glyn's conclusion be?

A His hypothesis is proved.

B His hypothesis is supported.

C His hypothesis is doubtful.

D His hypothesis should be rejected.

▶ How do scientists judge the strength of evidence?

To be confident that any conclusion you make is correct, you need strong evidence. Weak evidence does not mean your conclusion is wrong, it simply means you can be less sure it is right.

To judge the strength of evidence, you need to ask certain questions.

1 **How variable were the results?** The more variation in repeats, the weaker the evidence will be.

2 **Were enough repeats done? Was the sample big enough?** Variable results can still provide good evidence if the number of repeats or the sample size was big enough. You need to be certain that the results you have got were not 'freak' results. Freak results don't happen very often, so lots of repeats, or a big sample, mean that you will get a more accurate overall picture of what is happening.

3 **Were any differences significant?** Small differences may be just due to chance, as scientific measurements often cannot be perfectly accurate. Sometimes it is obvious that differences are or are not significant. If not, scientists may do statistical tests to measure how significant a difference is.

4 **Was the method flawed?** Faults in the method (for example, inaccurate ways of measuring, variables that could not be controlled) reduce the strength of the evidence. Major faults may mean that the conclusions are completely unreliable.

5 **Was the method valid?** A valid experiment is one that can provide an answer to the question it was investigating. For example, suppose you want to discover the effect of light intensity on the rate of photosynthesis. If you just move a lamp nearer and nearer to a plant and measure the rate, there is a problem. Light bulbs give off heat, and it could be that the heat was causing any changes, not the light. Unless you stop the temperature rise (for example, by shining the light through glass or water), then the experiment cannot answer the question and so is not valid.

⬇ Chapter summary

- Scientists investigate the world around them by a complex process of enquiry.
- Not all questions can be answered by science. A hypothesis is a suggested explanation for an observation that is based on evidence and can be tested by experiment.
- Evidence can either support or contradict a hypothesis, or may be inconclusive.
- A hypothesis is not a prediction, although it can be used to make predictions.
- A hypothesis can easily be disproved, but can rarely be proved.
- If a hypothesis is supported by lots of evidence and is generally accepted as true, it becomes a theory.
- To be of any use, an experiment must be fair and valid, and measurements must be as accurate as possible.
- Different measuring instruments have different levels of accuracy, linked to their resolution.
- If a variable cannot be controlled, the likely effect of not controlling it must be taken into account when analysing results.

- Repeating readings makes means more accurate, and allows repeatability to be assessed.
- The more variable the results are, the more times the experiment needs to be repeated (or the bigger the sample needs to be).
- Presenting data in tables makes it clearer than in text. The table should be constructed so that it is clear and the reader does not need to refer back to the method to see what it means.
- Line graphs and bar charts are used to make trends and patterns in the data clearer.
- Line graphs are used if both variables are continuous. Bar charts are used when the independent variable is discontinuous.
- The shape of a line graph indicates the nature and the strength of any trend or pattern.
- Evidence varies in strength. Stronger evidence makes the conclusion more certain.
- Experiments should also be reproducible – that is, give similar results each time the experiment is done, whoever it is done by.

Index

Acknowledgements

The Publisher would like to thank the following for permission to reproduce copyright material:

p.1 © Wellcome Library, London/http://creativecommons.org/licenses/by/4.0/; p.9 © Claude Nuridsany & Marie Perennou/Science Photo Library; p.12 © Cordelia Molloy/Science Photo library; p.17 © Michael Steele/Getty Images; p.19 © Biophoto Associates/Science Photo Library; p.21 © Ed Reschke/Photolibrary/Getty Images; p.24 t © Biophoto Associates/Science Photo Library, b © NIBSC/Science Photo Library; p.31 © Eye of Science/Science Photo Library; p.35 © egal/iStock/Thinkstock; p.42 © Carolina Biological/Visuals Unlimited/Corbis; p.58 © Power and Syred/Science Phot; p.59 l © Dr Jeremy Burgess/Science Photo Library, r © Tracy Tucker/iStock/Thinkstock; p.60 © Dr David Furness/Keele University/Science photo library; p.64 © swkunst/iStock/Thinkstock; p.71 © Rex Features/Eye Ubiquitous 26; p.74 l and m © British Lichen Society/Mike Sutcliffe, tr © Fotolia/Tatjana Gupalo, br © Irish Lichens/ Jenny Seawright; p.78 © Arterra Picture Library / Alamy Stock Photo; p.79 © MR1805/iStock/ Thinkstock; p.80 t © Lori Werhane/iStock/Thinkstock, bl © Goodshoot/Thinkstock, br © CathyDoi/iStock/Thinkstock; p.84 l © ImageBroker/Imagebroker/FLPA RF, r © Mark Sisson/ FLPA; p.86 © Martyn f. Chillmaid; p.87 l © Jupiter55/iStock/Thinkstock, r © Gary K Smith / Alamy Stock Photo; p.88 © The Photolibrary Wales / Alamy Stock Photo; p.89 © David Tipling Photo Library / Alamy Stock Photo; p.90 © Nigel Cattlin/FLPA; p.91 l © Nature Collection / Alamy Stock Photo, r © Imagestate Media; p.93 l © Photostock- Israel/Science Photo library, r © JohnatAPW/iStock/Thinkstock; p.94 © Steve Gschmeissner/Science Photo Library; p.95 © J. Craig Venter Institute; p.96 © Arkaprava Ghosh/Barcroft Media/Getty Images; p.97 © Steve Gschmeissner/Science Photo Library; p.103 © Science Photo Library; p.106 l © Christian Hütter / Alamy Stock Photo, r © Chris Burrows/Photolibrary/Getty Images; p.107 © Biophoto Associates/Science Photo Library; p.110 l © Photos by Sharon / Alamy Stock Photo, m © Pahham/iStock/Thinkstock, r © JoeGough/iStock/Thinkstock; p.112 © Imagestate Media Partners Limited – Impact Photos / Alamy Stock Photo; p.113 t © Gregg Porteous/ Newspix/ News Ltd, b © Asperra Images / Alamy Stock Photo; p.114 © S. Entressangle/ E.Daynes/Science Photo Library; p.117 © age fotostock / Alamy Stock Photo; p.118 l © Bain Collection/Library of Congress Prints and Photographic division, m © Gordon Chambers / Alamy Stock Photo, r © Paul Moore/Fotolia; p.119 © GL Archive / Alamy Stock Photo; p.122 t © Science Source/ Science Photo Library, b © Daily Mail Syndication/John Frost Newspapers; p.126 © Jamie_ Hall/iStock/Thinkstock; p.128 © FR Sport Photography / Alamy Stock Photo; p.135 © Cathlyn Melloan/Stone/Getty Images; p.136 © Alexander Raths/Fotolia; p.146 © BSIP SA / Alamy Stock Photo; p.151 © Satirus/iStock/Thinkstock; p.155 © Jake Lyell / Alamy Stock Photo; p.157 Oliver Strewe/Lonely Planet Images/Getty Images; p.158 © Ed Reschke/Photolibrary/ Getty Images; p.160 © Science Photo Library; p.167 © gbh007/iStock/Thinkstock; p.171 CALLALLOO CANDCY/Fotolia; p.173 © ELEN/Fotolia; p.177 © Jeffrey Banke/Fotolia.

t = top, b = bottom, l = left, r = right, m = middle

Every effort has been made to trace all copyright holders, but if any have been inadvertently overlooked, the Publisher will be pleased to make the necessary arrangements at the first opportunity.